essentials

essentials liefern aktuelles Wissen in konzentrierter Form. Die Essenz dessen, worauf es als „State-of-the-Art" in der gegenwärtigen Fachdiskussion oder in der Praxis ankommt. *essentials* informieren schnell, unkompliziert und verständlich

- als Einführung in ein aktuelles Thema aus Ihrem Fachgebiet
- als Einstieg in ein für Sie noch unbekanntes Themenfeld
- als Einblick, um zum Thema mitreden zu können

Die Bücher in elektronischer und gedruckter Form bringen das Fachwissen von Springerautor*innen kompakt zur Darstellung. Sie sind besonders für die Nutzung als eBook auf Tablet-PCs, eBook-Readern und Smartphones geeignet. *essentials* sind Wissensbausteine aus den Wirtschafts-, Sozial- und Geisteswissenschaften, aus Technik und Naturwissenschaften sowie aus Medizin, Psychologie und Gesundheitsberufen. Von renommierten Autor*innen aller Springer-Verlagsmarken.

Weitere Bände in der Reihe http://www.springer.com/series/13088

Bernd Herrmann · Bernhard Glaeser ·
Thomas Potthast

Humanökologie

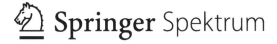
Springer Spektrum

Bernd Herrmann
Anthropologie, Universität Göttingen
Göttingen, Deutschland

Thomas Potthast
Lehrstuhl für Ethik, Theorie und
Geschichte der Biowissenschaften &
Ethikzentrum (IZEW)
Universität Tübingen
Tübingen, Deutschland

Bernhard Glaeser
Deutsche Gesellschaft für
Humanökologie, Freie Universität Berlin
& Wissenschaftszentrum Berlin für
Sozialforschung (WZB)
Berlin, Deutschland

ISSN 2197-6708   ISSN 2197-6716 (electronic)
essentials
ISBN 978-3-658-32982-2   ISBN 978-3-658-32983-9 (eBook)
https://doi.org/10.1007/978-3-658-32983-9

Die Deutsche Nationalbibliothek verzeichnet diese Publikation in der Deutschen Nationalbiblio-
grafie; detaillierte bibliografische Daten sind im Internet über http://dnb.d-nb.de abrufbar.

Springer Spektrum ist ein Imprint der eingetragenen Gesellschaft Springer Fachmedien Wiesbaden
GmbH und ist ein Teil von Springer Nature.
Die Anschrift der Gesellschaft ist: Abraham-Lincoln-Str. 46, 65189 Wiesbaden, Germany

Was sie in diesem *essential* finden können

- das *essential* vermittelt ein Grundverständnis von Menschen als Ergebnis einer biologischen Evolution, von Menschen als soziokulturell agierend und schließlich von Menschen als vernunftbegabten und moralfähigen Wesen
- das *essential* systematisiert die Hauptfelder der Humanökologie und führt zusammen

Vorwort

Humanökologie befasst sich mit dem Gesamt der Lebensansprüche der Art *Homo sapiens* gegenüber der Biosphäre. Sie befasst sich mit den ‚Dienstleistungs'-Ansprüchen sowie deren Umsetzungsfolgen und Nebenfolgen auf die von Menschen genutzten oder beeinflussten Habitate und Ökosysteme.

Drei Grundpositionen bestimmen dabei das *naturgeschichtlich inspirierte* Denken über Menschen: Menschen als Ergebnis einer biologischen Evolution, Menschen als soziokulturell agierende und schließlich Menschen als vernunftbegabte und moralfähige Wesen. An ihnen orientiert sich auch dieses *essential*.

Obwohl sämtliche Lebensäußerungen auf ihrer physischen Ausstattung und Befähigung gründen, ist eine Reduktion von Menschen allein auf den bloßen Zustand ihrer Biologie unangemessen und letztlich eine Fiktion. Menschen lediglich auf ihre organismische Ausstattung und Leistung zu reduzieren, so, wie sie selbst alle übrigen Lebewesen betrachten, würde wesentliche ihrer Eigenschaften und Lebensäußerungen völlig ausblenden.

Carl von Linné (1707–1778) hatte Menschen in sein System der Organismen *(Systema naturae)* nicht nur unter dem Gattungs- und Artnamen *Homo sapiens* aufgenommen. Das Epitheton „sapiens" (lat.: klug, vernünftig, weise) überschreitet eine Bewertungsgrenze, was nur aus ihrer Zeit (1758) zu verstehen ist. Für Linné war es zu Beginn der Aufklärung auch selbstverständlich, als ein differentialdiagnostisches Hauptkriterium für Menschen die lateinische Übersetzung der Aufschrift des Tempels des Apollon in Delphi zu übernehmen: „nosce te ipsum" (lat.: erkenne/kenne Dich selbst).

Nachdem Roger Bacon in seinem Werk *Novum Organum* (1620) die haupt-
sächlichen Vorurteile in wissenschaftlichem Gewande aufzählte[1], begründete er
damit die Forderung, auf Werturteile, auf ontologische Zuschreibungen bzw. Ver-
dinglichungsargumente zu verzichten. Diesen Anspruch verstärkte Bacon sogar
noch mit seiner Forderung „*Von uns selbst* [d. i. die Wissenschaftler[2]] *schweigen
wir.*"

Linné setzt sich bei der Systematisierung der ihm bekannten Menschenformen
über Bacons Mahnung zur Zurückhaltung hinweg und charakterisiert Menschen
unterschiedlicher geographischer Regionen in überheblicher bzw. herabsetzender
Weise. Selbst seine politischen Bewertungen finden ihren Niederschlag. Aus Frus-
tration über die Kriegsmüdigkeit seiner Schweden gegenüber den Russen bedient
Linné den Topos des *bon sauvage* mit einer späteren Verschiebung des Amerika-
ners in seiner Ordnung vor die seiner Meinung nach dekadenten, verweichlichten
Europäer. So sind es nach Linnés Meinung unterschiedliche Wesenszüge, welche die Menschen der damals bekannten Weltregionen charakterisierten, und sie
bestimmen entsprechend deren Platz in seiner Systematik: den *Homo sapiens
Americanus* zeichne eine feste rechtliche und moralische Ordnung aus, der *H.s.
Europäus* orientiere sich an Moden, der *H.s. Asiaticus* an subjektiven Einschätzun-
gen, und reine Willkür beherrsche den *H.s. Afer.* Man sollte nicht unterschätzen,
welche Auswirkungen derartige Bewertungsmuster, die sich als Vorbilder für wissenschaftlich daherkommende Beurteilungen anbieten, für spätere Auffassungen
von Menschen noch haben würden.

Linné war weder der erste noch der letzte, der eine derartige Grenzüber-
schreitung hinsichtlich des naturwissenschaftlichen Ideals beging. Die Verbindung
von Biologie und Politik bezüglich des Menschen lässt sich zurückverfolgen auf
die historisch frühe Bemühung von Aristoteles (384–322 BCE). Die vielzitierte
Aussage, wonach der Mensch ein politisches/staatliches Wesen wäre, findet sich
allerdings in einer politischen Schrift (Aristoteles, Politik 1. 1253a): „[…] *der
Mensch* [ist] *ein staatliches Wesen und zwar mehr, als die Bienen und die in Her-
den lebenden Tiere.*" Aristoteles weist damit auf die seiner Meinung nach den
Menschen innewohnende Bestimmung zur Bildung von Sozialgemeinschaften im
Sinne der seinerzeitlichen Polis, also einer Form der Staatlichkeit, hin.

---

[1]Francis Bacon erkannte vier Gruppen von Vorurteilen, die er Idola nannte [lat: idolum =
Götzenbild]: 1. Idola Tribus: Vorurteile, die in der menschlichen Natur selbst gründen, etwa
in der begrenzten sinnlichen Wahrnehmung; 2. Idola Specus: Vorurteile durch Erziehung,
durch Stimmungsschwankungen und mangelndes Wissen; 3. Idola Fori: Vorurteile durch
Selbstdarstellung, durch Kommunikationsprobleme und semantische Schwierigkeiten; Idola
Theatri: Vorurteile durch etablierte Lehrmeinungen und durch Schulenbildungen.

[2]Im gesamten Text wird das generische Maskulinum verwendet.

Offensichtlich stoßen Bestrebungen zur Ordnung des Naturganzen beim Gegenstand Mensch auf eine grundsätzliche Schwierigkeit. Ob nun aus dem Grund, dass Menschen, die über Menschen nachdenken, über den Gegenstand ihres Denkens viel mehr wissen als ihnen über jedes andere Lebewesen bekannt ist, oder aus der Einsicht, dass die Biologie allein zur Wesensbestimmung von Menschen von nur begrenzter Brauchbarkeit ist, immer bemühen die Denker metaphysische Weiterungen. Sie versuchen, eine ontologische Gewissheit, eine Letztursache für diejenigen menschlichen Eigenschaften anzugeben, die heute als die vorherrschenden oder bemerkenswertesten Eigenschaften gelten müssen. Werden diese Eigenschaften mit einem kreationistischen Ereignis (göttliche „Schöpfung") in Verbindung gebracht, erübrigt sich jedes Nachdenken darüber, ob die für Menschen charakteristisch gehaltenen Eigenschaften letztlich Folgen der auf Selbstorganisation der naturalen Prozesse beruhenden materiellen Welt sind.

Dabei tauchte die Idee eines von Gottheiten unbeeinflussten Naturganzen früh auf, Vertreter dieser Anschauung waren zum Beispiel Epikur (341 – 271/70 BCE) und Lukrez (99/94 – 55/53 BCE). Eine dieser Überzeugung nahestehende Einsicht David Humes (1711 – 1776) sollte auch für die moderne Naturwissenschaft leitend werden: „There is no Ought from an Is" („Es gibt kein Sollen aus dem Sein", d. i. deskriptive Aussagen können logisch keine normativen Aussagen fundieren; ein klassisch gewordenes Kriterium für den sogenannten Naturalistischen Fehlschluss). Doch dieses Humesche Gesetz konnte im Hinblick auf die Biosphäre nur solange gelten, wie deren Resilienz den lebenshinderlichen wie den mit dem Leben unvereinbaren Folgen menschlicher Handlungen abzupuffern in der Lage war. Längst geht es nicht mehr nur um Konservierung, sondern um die Sicherung einer wenigstens noch eingeschränkt Existenz bereitstellenden Biosphäre, so, wie wir sie heute noch kennen. Das „Sein" hängt mittlerweile völlig von einem einsichtsfähigen „Sollen" ab. Als Karl Marx formulierte: „Der Mensch tritt dem Naturstoff selbst als eine Naturmacht gegenüber" (Das Kapital, Bd I, Dritter Abschnitt), stand ihm noch das Bild des die Natur glänzend beherrschenden Ingenieurs, des Homo faber, vor Augen. Heute muss diese Fähigkeit allererst zur Abwendung der schädlichen Folgen und Nebenfolgen der auf den Naturstoff gerichteten menschlichen Handlungen eingesetzt werden.

Aus den kognitiven Eigenschaften von Menschen, aus ihrer „Begabung zur Vernunftfähigkeit" (Johann Gottfried Herder 1744–1803), resultieren Einsichtsfähigkeit und ethische Grundsätze, aus der Fähigkeit zur Folgenabschätzung resultieren Daseinsfürsorge und Vorsorgebefähigung. Ansätze dieser Eigenschaften sind auch bei anderen handlungsbegabten Organismen erkennbar. Bei keinem aber erlangen sie ein den Menschen vergleichbares qualitatives und quantitatives Niveau. Die Fähigkeit zur Umgestaltung ihrer Lebensräume bis hin zur existenzbedrohenden Beeinflussung des globalen Lebensraums sind offenbar Alleinstellungsmerkmale von Menschen. David Hume sah im unstillbaren Streben nach „Mehr" in allen Lebensbereichen (*„desire for gain"*) das charakteristische Merkmal menschlichen Daseins.

Als wirkliches Alleinstellungsmerkmal von Menschen wird sich die wiederholte Überschreitung der Grenzen des Evolutionsareals eines autochthonen afrikanischen Primaten erweisen, der in mehreren Auswanderungswellen schließlich alle Ränder des von Lebewesen besiedelbaren Teil der Erdoberfläche erreichte. Zweifellos lässt sich dies auf die kognitiven Fähigkeiten und technischen Fertigkeiten der Gattung zurückführen. Sie werden, wo auch immer Homo auftritt, zu einer sich selbst beschleunigenden Spirale des Erschließens neuer ökologischer Gegebenheiten, indem die neuen Habitate beziehungsweise Landschaften von menschlichen Bedürfnissen zunächst auf subtile Weise beeinflusst werden (sogen. *humanized landscapes*). Sie wurden in den nachfolgenden Zeiten der Sesshaftigkeit deutlich bis drastisch an den jeweiligen örtlich bedeutsamen kulturellen Vorgaben ausgerichtet.

Das Ergebnis ist bekannt. Aus der anfänglich bloßen Nutzung ökosystemarer Funktionen haben menschliche Sozialverbände und Gesellschaften durch andauernde Kolonisierungsmaßnahmen bis hin zu kaum standortgeeigneten Ausbeutungsstrategien einen gegenwärtigen Zustand der Biosphäre verursacht, der die Lebensfähigkeit der eigenen Spezies und die vieler anderer ernsthaft bedroht.

Zur Zukunftssicherung gehört ein breites Wissen über Wirkungen menschlichen Handelns auf die Natur. Hierzu will dieses *essential* beitragen, indem es Hintergrundwissen bereit stellt und für die Einsicht in die lokalen, regionalen und globalen Konsequenzen ökologisch wirksamen Handelns sensibilisiert. Es ist, mit Blick auf das Erdsystem und die Lage der Menschheit, überfällig, sich der „unsichtbaren Hand" (Adam Smith, 1723–1790), einer auf egoistische Ausbeutung zielenden Bewirtschaftung des Erdsystems, mit der sichtbaren Hand der ökologischen Vernunft entgegen zu stemmen.

Wir danken unserer Lektorin Stefanie Wolf und der Projektkoordinatorin Dagmar Kern, beide Springer Heidelberg, sowie Herrn Amose Stanislaus und Frau Omika Mohan, beide Chennai (Indien), Springer Nature, für die Betreuung des Buchprojektes.

BH dankt seiner Frau Susanne für Rat und Verständnis und den Mitautoren für die entspannte Zusammenarbeit, BG dankt den Kollegen BH und TP für

lehrreich-vergnügliche Auseinandersetzungen, seiner Frau Heide für das duld-
same Ertragen mentaler Abwesenheiten am Rechner, und TP dankt seiner Familie
für ihre Nachsicht sowie BH und BG für produktiv-humorvollen Austausch.

Wir hoffen auf eine günstige Aufnahme des *essential*.

Göttingen	Bernd Herrmann
Berlin	Bernhard Glaeser
Tübingen	Thomas Potthast
den 28. Februar 2021	

Inhaltsverzeichnis

Epistemologische Vorbemerkungen 1

Wissenschaftssystematisch ist die Humanökologie ein umfassender Wissenszusammenhang, der sich als Querschnittskonzept versteht und kaum auf ein einzelwissenschaftliches Verständnis reduziert werden kann. Erkenntnistheoretische Analysen im Bereich der Humanökologie lassen sich entsprechend methodisch nicht auf *einen* analytischen Zugang reduzieren. Gewiss wird man vorrangig die Mittel des Rationalismus verwenden, wobei sich Kompromisse zwischen den Positionen von Nicolai Hartmann (1882–1950), Karl Popper (1902–1994), Thomas Kuhn (1922–1996), Imre Lakatos (1922–1974) und Paul Feyerabend (1924–1994) anbieten. Selbstverständlich sind damit Überlegungen transtemporaler Vertreter rationaler Philosophie nicht obsolet. Die Humanökologie erinnert wegen ihrer Multithematik an (die letztlich vergeblichen) Versuche zur Institutionalisierung einer Gesamtwissenschaft Anthropologie, als deren Ergebnis am Ende des 19. Jahrhunderts einerseits die Biologie und andererseits die Geisteswissenschaften theoretische und institutionelle Durchsetzung erlangten (Frühwald 2006, S. 19–20) und sich die Einsicht durchsetzte, dass wissenschaftliche Wahrheit immer eine temporäre und relative ist. Tatsächlich scheint sich in der Humanökologie nahezu idealtypisch jene Kontextualisierung von Wissen zu realisieren (Elkana 2006)[1], in der deren Verknüpfung mit subjektiven Eindrücken zu dem führt, was individuell als „Weltzusammenhang" erkannt wird: „Nicht nur der Wissenschaft, sondern auch der Sprache, dem Mythos, der Kunst, der Religion ist es

---

[1]Elkana bezieht sich auf die von ihm sogenannte *Cassirerian contextualization of knowledge* (S. 127), einer Überprüfung (rethinking) des Wissensbestandes in der Tradition der Aufklärung. – Das auf Uexküll zurückgehende Zitat „Die Wissenschaft (oder Wahrheit) von heute ist der Irrtum von morgen" hat dieser allerdings nur sinngemäß formuliert (Uexküll 2014, S. 19).

eigen, daß sie die Bausteine liefern, aus denen sich für uns die Welt des ‚Wirklichen', wie des Geistigen, die Welt des Ich aufbaut." (Cassirer 2001, S. 22). Cassirer hat diese Einsicht unter anderem auch durch die Umwelttheorie seines Hamburger Kollegen Jakob von Uexküll gewonnen, wie er im 4. Kapitel seines „Versuchs über den Menschen" erläutert und fortführt:

> *Der Mensch kann der Wirklichkeit nicht mehr unmittelbar gegenübertreten; er kann sie nicht mehr als direktes Gegenüber betrachten. Die physische Realität scheint in dem Maße zurückzutreten, wie die Symboltätigkeit des Menschen an Raum gewinnt. Statt mit den Dingen hat es der Mensch nun gleichsam ständig mit sich selbst zu tun. So sehr hat er sich mit sprachlichen Formen, künstlerischen Bildern, mythischen Symbolen oder religiösen Riten umgeben, daß er nichts sehen oder erkennen kann, ohne daß sich dieses artifizielle Medium zwischen ihn und die Wirklichkeit schöbe.* (Cassirer 1996, S. 50).

Uexküll hatte, möglicherweise in Fortsetzung des Raumkonzeptes des Biologen und Geographen Friedrich Ratzel (1844–1904), die Idee eines jeden Lebewesens allein zukommenden Raumes um Komponenten der selektiven Wahrnehmung und Wirkung des einzelnen Organismus erweitert. In „Umwelt und Innenwelt der Tiere" (1909/1921) und nachfolgend erläuternden Publikationen fand er hierfür die Metapher eines Lebens wie unter einer gläsernen Glocke bzw. in einer Seifenblase, das er mit dem Begriff „Umwelt" belegte (Herrmann 2021). Ob dabei die kurz vor Uexkülls Erstauflage 1909 in deutscher Sprache erschienenen Überlegungen des amerikanischen Philosophen William James (1842–1910) berücksichtigt wurden, ist bisher nicht verfolgt worden. Verblüffend sind jedenfalls die Übereinstimmungen mit Uexkülls Einsichten wie auch denen Ernst Cassirers, die sich in einer Kernaussage von James (2012, S. 151) wiederfinden: „Was wir über die Wirklichkeit aussagen, hängt also von der Perspektive ab, aus der wir sie betrachten. *Dass* die Wirklichkeit existiert, können wir nicht beeinflussen; aber *was* sie ist, beruht auf einer *Auswahl*, und diese Auswahl treffen wir."[2] Zugleich ist damit das grundsätzliche Dilemma beschrieben, dass die Wirklichkeit, die Totalität alles Existierenden („die Natur"), von Menschen nur aus der Erste-Person-Perspektive beschrieben werden kann. Der Anthropozentrismus resp. das sogen. allgemeine anthroporelationale Prinzip ist unhintergehbar und die Behauptung der (Natur)Wissenschaften, ein *objektives* Bild der Natur und der Wirklichkeit zeichnen zu können, gilt nur innerhalb von Verabredungen auf

---

[2]Es ergibt sich ein unmittelbarer Anschluss an das ‚Thomas-Theorem' der Soziologie (nach W.I. Thomas 1928): „If men define situations as real, they are real in their consequences", nach dem die objektive Situation und ihre subjektive Bewertung zusammen fallen können.

bestimmte Standards der Wirklichkeitsbeobachtung und Wirklichkeitserklärung (vgl. Vorwort, Fußnote 1).

Die Wirklichkeit ist untrennbar mit der Frage verbunden, *was* denn „*Natur*" sei. Die erste Annäherung als „die Totalität alles Existierenden" (Hans-Dieter Mutschler), führt sofort in den gedanklichen Zweifel, ob die neben der sich selbst hervorbringenden Natur *(natura naturans)* existierenden Artefakte menschlicher wie tierlicher Herkunft mit eingeschlossen sind. Neben der Natur „als Wesen einer Sache" *(natura prima)* kann das, was einem Lebewesen im Laufe seines Lebens mitgegeben wird, zur *natura secunda* werden, wie etwa in dem berühmten Beispiel der Bärin Ovids (Metamorphosen: 15. Buch, 147), die ihr Junges in Form leckt (zu den unterschiedlichen Naturbegriffen Herrmann 2016, S. 247 ff.). *In diesem Buch wird zwischen den drei genannten Naturbegriffen in Bezug auf den Menschen nicht unterschieden. Seine kulturellen Leistungen werden seiner Natur zugerechnet.* Die Trennung von Natur und Kultur gilt nach heutiger Einsicht als überholt, weil sie lediglich eine von vielen Möglichkeiten ist, die Totalität des Existierenden einzuteilen (Descola 2013). Sie hat innerhalb unseres logischen Systems eine gewisse heuristische Brauchbarkeit, stellt aber als ausschließende binäre Trennung einen Kategorienfehler dar, ähnlich jenem von Körper und Geist (Herrmann 2016, S. 39 ff.).

An dieser Stelle öffnet sich dann auch zwangsläufig die Eintrittspforte für ontologische Zuschreibungen, für wertebasierte Entscheidungen und handlungsleitende Überzeugungen auf die Wirklichkeit, für Konzepte, wie den ökologischen Imperativ: „Handle so, dass die Wirkungen deiner Handlung verträglich sind mit der Permanenz echten menschlichen Lebens auf der Erde" (Hans Jonas, 1903–1993), der in den Zeiten des Biodiversitätsverlustes eine Ergänzung erfahren muss: „Handle so, dass die Wirkungen deiner Handlung verträglich sind mit der Permanenz *allen* Lebens auf der Erde". Dieser normativen Forderung entspricht in der Bundesrepublik Deutschland u. a. GG Art 20a: „Der Staat schützt auch in Verantwortung für die künftigen Generationen die natürlichen Lebensgrundlagen und die Tiere im Rahmen der verfassungsmäßigen Ordnung durch die Gesetzgebung und nach Maßgabe von Gesetz und Recht durch die vollziehende Gewalt und die Rechtsprechung." Diese Forderung findet eine verständliche Nichtanwendung bei Kalamitäten, gegenüber Krankheitserregern und Parasiten, und sie ist als nationales Gesetz auf den Geltungsbereich des Grundgesetzes begrenzt. Die allgemeine, auch überregionale Zumessung eines Wertes für ein Lebewesen jenseits bloßer Gründe des menschlichen Lebenserhalts oder im Interesse unmittelbarer wirtschaftlicher Verwertbarkeit ist das Kernstück ethischer Erwägungen, die im Kap. 4 dieses Essentials behandelt werden.

In der Humanökologie können Erkenntnismittel der strengen Naturgesetzlich-keit wissenschaftssystematisch nicht bzw. nur segmental angewendet werden. In ihr ist der Zusammenhang wissenschaftlicher Aussagen hinsichtlich Aussages-ensitivität und Aussagespezifität sozionaturaler Abläufe und Gegebenheiten zu beachten. Die Relevanz wissenschaftlicher Aussagen nimmt mit steigender Aus-sagepräzision ab: eine Aussage, Feststellung, Bewertung trifft auf umso weniger Fälle zu, je konkreter die berücksichtigten Randbedingungen werden (d. i. bei zunehmender Spezifität der Determinanten). Dies gilt in besonderer Weise für das sozialpsychologische Verhalten und Handeln von Menschen. Dennoch ist zu beachten, dass am Ende jedes Verhaltens, jedes noch so komplexen Gedan-kens wie auch jeder Wahrnehmungstäuschung materielle Prozesse stehen, atomare Abläufe, die einer Naturgesetzlichkeit folgen. Selbst wenn man aus Gründen „spezifischer Komplizierung" (Nicolai Hartmann) noch Zuflucht beim Zufall suchen möchte, so folgt auch dieser der allgemeinen Naturgesetzlichkeit: „Al-les, was einmalig geschieht, steht gleichwohl in allen seinen Einzelbestimmungen unter allgemeinen Prinzipien. Es ist also nichtsdestoweniger auch die Wesens-notwendigkeit des streng Allgemeinen an sich. Dieses streng Allgemeine ist die Naturgesetzlichkeit" (Hartmann 1980, S. 399–400). Die Naturgesetzlichkeit ergibt sich in der retrospektiven Analyse, weil der Zufall nicht planbar ist, aber die Wege zu seinem Ereignis logisch nachvollziehbar sind bzw. sein würden (vgl. Herrmann und Sieglerschmidt 2018, S. 18). Damit steht eine materialistische Betrachtung im Zentrum der Humanökologie.

Hinzuweisen ist noch darauf, dass im Folgenden keine Beschreibung der öko-logischen Situation des Weltsystem und seiner Kompartimente gegeben wird. Hierzu wird auf die Zustandsberichte der UN und ihren Gruppierungen ver-wiesen (z. B. FAO 2007, 2010, 2020; MEA 2005; IPBES 2019; IPCC 2014). Konkrete Hinweise werden nur stellvertretend und ihrer Beispielhaftigkeit wegen aufgeführt, wenn sie dem erleichternden Verständnis dienen.

Ökologisch-organismische Grundlagen

2

In ökologischen Prozessen geht es *immer* um die Balance zwischen den sechs ökologischen Grundelementen: *Stoff* (d. i. Materie als Voraussetzung), *Energie* (für die Aufrechterhaltung der lebensnotwendigen Prozesse), *Raum* (als Aufenthaltsort für Lebewesen), *Information* (genetische Programme in den Lebewesen und akkumuliertes Wissen („tradigenetisch")), *Zeit* (als immanente Prozessgröße) und *Biota* (die Lebewesen im konkreten Raum). Ökologische Faktoren, d. i. das Wechselspiel zwischen Umweltfaktoren und der organismischen Anpassung, waren, neben genetischer Drift, die Ursachen der evolutiven Differenzierung auch für die Menschwerdung – jedenfalls solange, wie sich dieser Prozess im Zustand der selbstlaufenden Biologie abspielte. Wie alle Lebewesen variieren Menschen in ihren körperlichen Eigenschaften, je nach Geburtsort und Abstammungsgemeinschaft. Individuelle Variabilität der Lebewesen ist die Versicherung einer biologischen Art bzw. einer Population für ihren weiteren Bestand, gegen ökosystemare Änderungen, Störfaktoren und lebenshinderliche Zufälle. Menschliche Diversität ist eine Folge evolutiver Prozesse, eine Folge von Anpassung an geographisch variierende physikalische, chemische und biologische Determinanten. Nach Maßstäben kosmopolitischer Verbreitung und Individuenzahlen war die Diversitätsbildung die Grundlage für den Erfolg der Art *Homo sapiens*.

2.1 Ursprung und Verteilung der Menschheit

Die Faktoren, die zur Entwicklung von Menschen führten, werden mit einem miozänen Umbruch (vor ca. 6,7 – 6,2 Mio a) in den Klima- und Umweltverhältnissen im afrikanischen Ursprungsraum, der Savannenlandschaft, gesehen. Mit

der klimatischen Veränderung traten ausgeprägte saisonale Umweltschwankun-
gen auf. Entsprechend wird in der ökologischen Ursache „Saisonalität" mit ihren
Merkmalen wie Temperaturschwankungen, Habitatmosaiken und Nahrungsver-
fügbarkeit der Auslöser wie Hauptreiber gesehen. Für die spezifisch menschliche
Evolutionsgeschichte soll die „Theorie der optimalen Nahrungssuche" (opti-
mal foraging theory; Schutkowski 2006, S. 62–67) eine herausgehobene Rolle
gespielt haben. Sie verrechnet die energetischen Kosten der Nahrungsbeschaffung
und -zubereitung gegen die energetischen Kosten des nutzbaren Nahrungsanteils
und vergleicht mit einer rein opportunistischen Nahrungsaneignung. Zugleich
soll in dieser Nahrungsbeschaffungsstrategie die Wurzel für Mechanismen der
menschlichen Entscheidungsfindung zu sehen sein. Weiter erklärt die Theorie geschlechtsspezifische Rollenverteilungen, zunächst der Nahrungsbeschaffung
(Jagen durch Männer vs. Sammeln durch Frauen).[1] Damit verbunden soll die
Bevorzugung erfolgreicher Jäger durch kopulationsbereite Frauen gewesen sein
und die schließliche Entstehung des Daueröstrus bei gleichzeitig verdeckter
Ovulation, die primatenuntypisch ist.

Die evolutionsökologischen Konzepte behaupten weitreichende Konsequenzen.
So soll Bipedie den Transport von Jagdbeute begünstigt haben. Der kalorische
Ertrag erlegter Jagdbeute solle Zeit für die Herstellung von Gerätschaften und
damit Werkzeugherstellung und die manipulativen Fähigkeiten der Hand geför-
dert haben. Schließlich hätte es ein höheres Investment der Eltern in ihre Kinder
gegeben. Die Frauen hätten durch Zuteilung von hochwertiger Nahrung sich mehr
um die Kinder kümmern können, und die Versorgung älterer Gruppenmitglieder
(Bewahrung des Erfahrungsschatzes) wäre gesichert worden.

Neben diesem Modell der Ernährungsstrategie, existiert ein alternatives,
wonach die Fleischbeschaffung vorrangig über das Stehlen von Jagdbeute von
Raubtieren bewerkstelligt wurde (Aasesser-Modell).

Die evolutionsökologischen Konzepte unterstellen einheitlich, dass Bipedie
und die Entwicklung einer materiellen Kultur die adäquate Anpassungsstrate-
gie für eine optimale Ressourcennutzung wären (zusammengefasst aus Henke
und Rothe 2014, dort auch Angaben zum zeitlichen Auftreten früher Menschenformen, aus deren Formenkreis die Entwicklungslinie zur Gattung Homo
führte).

---

[1]Diese Theorie ist in die wissenschaftliche Kritik geraten.

Out of Africa

Die heute weltweit verbreitete Menschheit, der ‚anatomisch moderne Mensch‘ *(Homo sapiens sapiens)*, hat einen gemeinsamen, *monophyletischen* afrikanischen Ursprung (Cavalli-Sforza et al. 1994; Henke und Rothe 2014; Reich 2018).

Ihre Verbreitung beruht auf mehreren Auswanderungswellen aus der afrikanischen Urheimat. Normalerweise verlassen Arten den Raum ihrer Entstehung nicht, weil die ökosystemaren Bedingungen ihrer Existenz durch die Grenzen des spezifischen Ökosystems vorgegeben sind. Über Ursachen, die letztlich zur Auswanderung führten, wird spekuliert. Möglicherweise beruhen die Auswanderungswellen auf einer frühen Bevölkerungszunahme. Voraussetzung für die Ausbreitung waren auch, dass zeitgenössische Ausbreitungsbarrieren u. a. durch klimatische Gunst überwunden werden konnten (leichtere Erreichbarkeit u. a. durch niedrige Meeresspiegel) und die flexible Anpassung an die Ressourcenverfügbarkeit in den außerafrikanischen Biomen. Diese unterscheiden sich biogeographisch grundsätzlich vom afrikanischen Ursprungsraum, als dieser zwischen den Wendekreisen des Nord-Süd-ausgerichteten afrikanischen Kontinents liegt, die Ausbreitung aber hauptsächlich durch Zonen oberhalb des nördlichen Wendekreises im West-Ost-ausgestreckten eurasischen Verbundkontinent erfolgte. Menschen erwiesen sich als Habitatsgeneralisten. Möglich war die Ausbreitung durch unterschiedlichste Lebensräume und in diese hinein nur auf einer Grundlage soziokultureller und technologischer Kompetenzen und durch die Bereitschaft zu explorativer Umweltaneignung.

In anatomisch modernen Menschen sind die Vorfahren der letzten beiden Auswanderungswellen vereint. Aus der vorletzten Welle (vor ca. 600.000 Jahren) entwickelte sich u. a. der *Homo neanderthalensis*. Er hinterließ einen kleinen Anteil seines Genmaterials in den Menschen der etwa bis zu 200.000 Jahre zurückliegenden letzten Auswanderungswelle, die heute als *Homo sapiens sapiens* die Erde besiedelt haben.

Die gegenwärtige globale Diversität von Menschen ist also jünger als 200.000 Jahre. In einigen Regionen der Erde ist sie sogar gerade einmal einige zehntausend Jahre (z. B. N- und S-Amerika) bzw. nur einige tausend Jahre alt (z. B. Ozeanien). Aus genetischen Daten lassen sich die Ausbreitungswege der letzten 150.000 Jahre rekonstruieren (Reich 2018).

2.2 Ökologische Parameter

Menschen sind Nahrungsgeneralisten, eine wesentliche Voraussetzung für das Migrationsverhalten und die Besiedlung unterschiedlichster Biome. Ökologisch ist die Häufigkeit von Generalisten umgekehrt mit der Produktivität eines Habitats korreliert: mit abnehmendem Nahrungsangebot bevorzugt der Selektionsdruck generalistische Verwerter. Menschen leben heute in Gegenden, deren mittlere Jahresdurchschnittstemperatur zwischen −17 °C und +38 °C liegt. Sie leben in Höhenstufen zwischen dem Meeresspiegel und 5500 m. Die Siedlungsaktivität fällt oberhalb 4000 m und nahe den Temperaturextrema deutlich ab.

Das breite Spektrum von Umweltfaktoren erfordert physiologische Anpassungen der Wärmeregulation im Stoffwechsel und Kreislaufsystem. Es gibt physiologische Spezialisierungen und anatomisch adaptive Eigenschaften, die auf speziellen Genotypen beruhen sowie spezifische kulturelle und soziale Anpassungen. Menschen kommen in vielen verschiedenartigen Lebensräumen vor, sie sind *eurytop*. Menschen können Schwankungen lebenswichtiger Umweltfaktoren innerhalb weiter Grenzen ertragen, sie sind *euryök*. Wobei ungünstige Umweltparameter bei zunehmender Abträglichkeit mittels kulturell erworbener Mittel in Grenzen kompensiert werden können. Es hängt von der Betrachtungsweise ab, ob Menschen als im ökologischen Sinne euryök gelten sollen, je nachdem man die kulturell erworbenen Kompetenzen ‚der Natur des Menschen' zurechnet oder nicht. In diesem Buch werden die kulturellen Erwerbungen der Menschen ihrer Natur zugerechnet.

Menschen verdanken ihr äußerliches Erscheinungsbild vor allem thermoregulatorischen Anpassungen, von der Haarform über die Hautfarbe bis hin zum Körperbautypus. Wie das phänotypische Erscheinungsbild basieren auch die physiologischen Eigenschaften von Menschen auf genetischen Grundlagen. Die genetischen Eigenschaften von Menschen weisen klare geografische Gradienten als Anpassungsfolgen auf (Cavalli-Sforza et al. 1994). Ihre Verteilungsunterschiede gehen v. a. auf Anpassungen an biogeografische Gegebenheiten, das Vorkommen von Parasiten und Krankheitserregern zurück. Dabei spielen auch Koevolutionen zwischen kulturellen Spezifika menschlicher Fortpflanzungsgemeinschaften, ihren genetischen Ausstattungen und naturräumlichen Determinanten eine erhebliche Rolle (Durham 1991; Moran 2008).

2.3 Das menschliche Ökosystem

Prinzipien des Ökosystems

Grundsätzlich ist der Unterhalt der Lebensfunktionen an die Verfügbarkeit und den Zugang zu den energetischen und stofflichen Grundlagen für die Lebensprozesse einschließlich der Reproduktion gebunden. Die Reproduktion ist das Mittel zur Verstetigung, zur Permanenz des Lebensprozesses. Allerdings konkurriert das

Streben nach Permanenz mit demselben Streben anderer Organismen, denn die verfügbare energetische und stoffliche Grundlage ist begrenzt. Der Zugang zu den Ressourcen wird durch Aushandlungsprozesse zwischen den biologischen Akteuren geregelt. Es handelt sich dabei um Konkurrenzverhalten zwischen den Individuen einer Lebensgemeinschaft. Die Idee einer ökosystemaren Dienstleistung (s. u.) geht von der Gewährleistung der Verfügbarkeit der energetischen, der stofflichen und reproduktionspartnerschaftlichen Grundlagen aus. Wenn dabei die Handlungen bzw. Beziehungen zwischen den Individuen so ausbalanciert verlaufen, dass eine für die erfolgreiche Reproduktion notwendige Anzahl von Individuen aller Arten eine gesicherte Existenzgrundlage findet, befindet sich diese Lebensgemeinschaft in einem stabilen Gleichgewicht, dem Zustand der „Nachhaltigkeit". Nachhaltigkeit bezeichnet die Bedingungen, unter denen die lebenserhaltenden Prozesse oder Bedürfnisdeckungen verstetigt werden können. Real beeinflussen Zufälle wie Weidegänger, Prädatoren, Krankheitserreger, Parasiten, Mutationen, Unfälle, Schwankungen in den Umweltmedien oder Extremereignisse den Verlauf der Aushandlungsprozesse und beeinflussen die Stoff- und Energieflüsse im Ökosystem. In der Regel können Normalitätsschwankungen durch die Pufferkapazität des Systems aufgefangen bzw. ausgeglichen werden *(Resilienz),* was prekäre Zustände für einzelne Individuen oder Arten nicht ausschließt. Ob sich dabei der Gesamtcharakter des Systems ändert, ist eine Frage skalenabhängiger Beurteilung.

Jedes Lebewesen nimmt einen physischen Raum in einer Umgebung ein, der durch die körperliche Befähigung des Lebewesens und die für seine Lebensfähigkeit verfügbaren Umweltdeterminanten begrenzt ist. Als Denkmodell für den Existenzraum hat die Biologie die Idee des Ökosystems entwickelt. *Ein Ökosystem kann verstanden werden als ein durch Selbstorganisation der Wirkungen einer endlichen, wenn auch nicht notwendig bekannten, Anzahl von Arten und Umweltmedien aufeinander in einem bestimmbaren geografischen Raum entstandene raumzeitliche Gemeinschaft von Lebewesen und ihren medialen Substraten.* Das Ökosystem ist als Strukturganzes der *logische Raum* sämtlicher ökosystemarer und auf sie gerichteter ideenmäßiger Sachverhalte. Das System unterliegt in seiner Entwicklung Zeitformen (Kirchhoff 2015), ohne dabei seine grundlegende Charakteristik zu ändern.

Das Apriori eines Lebewesens liegt in der Deckung seiner Lebensansprüche. Die Summe der Lebensansprüche und ihr Erfüllungsraum bilden das spezifische ökologische Anspruchsgefüge eines jeden Lebewesens. Ausschließlich unter den Bedingungen des Systems ist eine Verstetigung der lebensbegründenden Prozesse und der Bedürfnisdeckungen möglich. Diese Verstetigung ist das intrinsische Prinzip, auf das alle Lebensvorgänge ausgerichtet sind und das die eigentliche und grundlegende Nachhaltigkeit aller Lebensprozesse darstellt. Umgekehrt verdankt sich das Ökosystem selbst nur der Tatsache, dass seine Lebewesen in ihm verstetigt existieren. Das System bringt sich selbst hervor.[2]

Ökosystemare Dienstleistung
In der anthropozentrischen wie anthroporelationalen Betrachtung der von Menschen belebten Ökosysteme wird die Anspruchserwartung formuliert, dass die Systeme *Dienstleistungen (ecosystem services)* zur Verfügung stellten(z. B. MEA 2005 bzw. TEEB 2008; Lemma *Ökosystemdienste* in Schaefer 2012; IPBES 2019). Der Begriff ist eine terminologische und didaktische Entgleisung. Seine anthropozentrisch-egoistische Position erklärt zumindest einen Teil des heutigen allgemeinen Umweltdilemmas. Er schließt außerdem an Denkfiguren an, die in vielen Überzeugungssystemen und Religionen die Erschaffung der Welt zum Nutzen der Menschen behaupten. Für die Einsicht der Naturwissenschaften, wonach das Gesamtprodukt Welt das Ergebnis selbstorganisierender Prozesse ist, beruht ein solcher Rückgriff auf fiktionalen Bedeutungszuschreibungen.

Kultur als menschliches Ökosystem
Im Falle von Menschen erfolgt die Deckung der individuellen wie kollektiven Lebensansprüche faktisch wie normativ über Bedeutungszuschreibungen, die durch soziale Interaktionen und damit verbundene interpretative Prozesse innerhalb eines Sozialverbandes entstehen. Grundlegend ist hierbei der sozial geregelte Umgang mit den lebensbegründenden Ressourcen und Realien. Max Webers Definition der Kultur als *„ein vom Standpunkt des Menschen aus mit Sinn und Bedeutung bedachter endlicher Ausschnitt aus der sinnlosen Unendlichkeit des Weltgeschehens"* (Weber 1988, S. 180) gilt entsprechend für das spezifisch menschliche Ökosystem. In der Substanz ist „Kultur" der mit Lebewesen und Umweltelementen ausgestattete physische, soziale und philosophische Raum, in dem Menschen sich bewegen und der ihnen mit seinen Dienstleistungen das Überleben sichert: *Die Kultur* ist *das menschliche Ökosystem* (Herrmann 2019).

Jede Kultur versucht mit den Möglichkeiten ihrer Konzeption, allermeist unterstützt durch Sprachgrenzen, die Existenz ihrer Angehörigen nachhaltig zu sichern.

---

[2]Anders ist dies bei den anthropogenen Ökosystemen, etwa der Landwirtschaft, die der beständigen Kolonisierungsarbeit bedürfen, weil sie sonst durch nichtanthropogene ökosystemare Sukzessionen ersetzt würden. Seitdem ist „Arbeit" als ein Zentrum menschlichen Selbstverständnisses etabliert. Sie verdrängt(e) mit ihrem Verhaltensleitbild Produktivität und der damit verknüpften Idee der Vorratsgesellschaft die ökologieverträglichen Säulen einer steinzeitlichen Lebenseinstellung von Unterproduktivität und Mußepräferenz (Sahlins 1972; Groh 1992).

Kulturen ‚managen' Bedürfnisse und Dienstleistungen sowohl ganzer Bevölkerungen als auch ihrer Teile (Subkulturen). Im historischen Prozess der allmählich weltweiten Ausbreitung der anatomisch modernen Menschen verbirgt sich die Geschichte der ökologische Wirkungen der menschlichen Kulturen als quasi Suprastrukturen bzw. managenden Leitideen über die von ihnen genutzten bzw. besiedelten Ökosysteme (Isenberg 2017; McNeill 2005; Herrmann 2016; Robinson und Wiegand 2008).

Durch den existenzbegründenden Umweltbezug eines jeden Lebewesens ergibt sich zwangsläufig umgekehrt eine Beeinflussung *(impact)* der Umwelt. Da Menschen nicht kulturfrei gedacht werden können, ist ihr Umweltbezug immer abhängig von spirituellen, normativen und institutionalisierten Vorstellungen bzw. Handlungsanleitungen, die ihren Niederschlag in Wirkungen auf ein Ökosystem finden. Die Gesamtheit der durch Menschen bedingten Einflüsse auf Lebensräume und Lebensgemeinschaften wird im ökologischen Begriff *Hemerobie* [ἥμερος = kultiviert (Ackerboden), gezähmt] zusammen gefasst. Der Grad der Beeinflussung ist nur in relativer Begrifflichkeit zwischen *ahemerob* (unbeeinflusst), *euhemerob* (stark beeinflusst) und *metahemerob* (vernichtet) und Zwischenstufen anzugeben.

Für die Selektionsprozesse bei gegenwärtigen anatomisch modernen Menschen haben biologische Faktoren nicht mehr jene herausgehobene Bedeutung im Verhältnis zu den vorrangigen kulturellen, institutionalisierten Erwerbungen. Kulturell adaptive Strategien haben u. a. ein Überleben in pathogen belasteten Regionen möglich gemacht (Durham 1991). Technische Fertigkeiten, sozial gesicherte Allokation von Ressourcen sowie ethische Standards können gegenwärtig in vielen menschlichen Kulturen das Überleben sichern, mitunter auch den reproduktiven Erfolg von Individuen, deren Prognose im reinen Zustand der Biologie infaust wäre.

2.4 Bevölkerungen

Von Beginn an zeigten Bevölkerungen anatomisch moderner Menschen ein ausgeprägtes Migrationsverhalten, das bis auf den heutigen Tag anhält. Als Hintergründe spielen Verdrängungen durch expansive Bevölkerungsgruppen ebenso eine Rolle wie ethnische Konflikte, Desertifikationen und Ernteausfälle, Verwüstungen durch kriegerische Handlungen oder naturale Extremereignisse, ökonomische Perspektivlosigkeit. Alle diese Gründe beruhen auf einer Nichterfüllung der erwarteten Dienstleistungen des menschlichen Ökosystems.

Ein gravierendes Proxidatum für den heute prekären ökologischen Zustand des Weltsystems ist die Entwicklung der Weltbevölkerung: Im Grundsatz kann die

Agrikultur gegenüber Wildbeutertum und pastoralem Nomadismus die energetisch ertragreichere Wirtschaftsform sein, sie wettet auf höhere pro-Kopf-Erträge und kontinuierlich Versorgung. Allerdings waren Hungersnöte ein weltweiter ständiger Begleiter der Agrarsysteme (Pathogene, Schädlinge, klimatische Ungunst, Extremereignisse) und damit ein prosperitätsbegrenzender Faktor. Erst die Möglichkeiten zur Substitution der Bodenfruchtbarkeit, die dauerhafte Lösung überregionaler Transport- und Verteilungsprobleme und die Einführung hygienischer Mindeststandards erlaubte im 20. Jahrhundert den weltweiten Anstieg der Lebenserwartung und einen dramatischen Anstieg der Bevölkerungszahl.

Das verhaltensökologische Grundproblem menschlicher Bevölkerungen ist, dass Menschen als Primaten eigentlich biologische *K-Strategen* sind (wenig Nachkommen, in die stark investiert wird, um in besetzten Nischen wettbewerbsfähig zu sein), die sich aber aus vorrangig sozialen Gründen wie *r-Strategen* verhalten [viele Nachkommen bei geringer individueller Überlebenswahrscheinlichkeit, um (unbesetzte) Nischen auszubeuten bzw. als elterliche Altersversorgung]. Erst mit dem Erreichen eines relativen Bildungsniveaus und Lebensstandards[3] trat der sogenannte Demographische Übergang ein: die Ablösung einer hohen Geburtenrate und hohen Sterblichkeit durch eine zuerst geringere Sterblichkeit, gefolgt von einer abnehmenden Geburtenrate. In Deutschland begann der Demographische Wandel nach 1870, wobei gegenwärtig die Zahl der Sterbefälle die der Geburten überwiegt.

Die UN schätzt die Zahl der Weltbevölkerung um das Jahr 0 CE auf 300 Mio. (= 1), im Jahre 1900 auf 1,65 Mrd. (= 5,5x), im Jahre 2020 auf 7,8 Mrd. (=26x). Mit den Neolithischen Revolutionen sank die durchschnittliche Lebenserwartung von 30 Jahren (Wildbeuter) auf zunächst 20 Jahre, wegen eines zunehmenden Lebensrisikos (Arbeitsbelastung, unsichere Versorgungslagen, Bevölkerungsdichte, Infektionsrisiken, Zoonosen). Sie stieg bis ins 19. Jahrhundert auf ca. 40 Jahre (Deutsches Reich 1871: ♂♂ 39 Jahre ♀♀ 42 Jahre), also eine annähernde Verdoppelung der Lebenserwartung in 10.000 Jahren. Demgegenüber beträgt sie heute in Deutschland mit rund 80 Jahren das Vierfache der neolithischen Lebenserwartung zum Zeitpunkt der Geburt. Für die ehemals geringe Lebenserwartung war überwiegend die durchschnittlich hohe Kindersterblichkeit verantwortlich.

---

[3]Das sogen. ökonomisch-demographische Paradoxon: eine Bevölkerung oder soziale Schicht bekommt umso weniger Nachwuchs, je gebildeter und wohlhabender sie ist. In Europa setzte dieses Verhalten zuerst beim britischen Adel ab dem 17. Jahrhundert ein, ohne jedoch allgemeine Vorbildfunktion auszuüben.

Die Bevölkerungsstruktur reflektiert vielfältig ökologische bzw. sozialökologische Parameter einer Bevölkerung. Die Individuen einer Alterskohorte bilden nach Geschlechtern getrennt das Grundgerüst einer Bevölkerungspyramide, die bevölkerungsrelevante Großereignisse (Epidemien, kriegsbedingte Verluste, Schwankungen der Fertilität) unmittelbar abbildet. Geringer aggregierte Datensätze können nach zahlreichen Sozialgruppen differenzierte Lebenserwartungen und Auswirkungen sozialökologischer Faktoren erfassen.

Der heutige Zustand des Erdsystems ist ein Ergebnis ‚der Geschichte' als des Gesamt der von der Weltbevölkerung verursachten Zustände und Wirkungen. Deren Anteile müssen im Hinblick auf den gegenwärtigen Zustand des Erdsystems als *hinreichende* Bedingung gelten, während die übrigen Lebewesen und Naturkreisläufe lediglich als *notwendige* Bedingungen für das basale Tableau sorgten. Die Lebensansprüche der heutigen Weltbevölkerung sind der eigentliche Grund für dessen gegenwärtig desaströsen Zustand. In den letzten 70 Jahren haben Menschen das Ökosystem der gesamten Erde schneller und umfassender verändert als in irgendeinem davor liegendem Zeitraum, im Wesentlichen, um den rasch anwachsenden Bedarf an Nahrung, Wasser, Holz, Fasern und Treibstoff zu decken, was gleichzeitig mit einem substantiellen Biodiversitätsverlust verbunden ist (MEA 2005).

2.5   Die Aneignung natürlicher Ressourcen

Aus ökologischer Perspektive ist die für menschlicher Angelegenheiten vorgenommene Trennung von Ökologie und Ökonomie nicht sinnvoll, weil die Ökonomie hier als kulturenspezifische Aneignung der ökologischen Grundkategorien und natürlichen Ressourcen zu verstehen ist. Im gängigen Fortschrittsmodell der menschlichen Weltaneignung führen die *Neolithischen Revolutionen* (um ca. 11.000 BCE und jünger) zu nahrungsproduzierenden Wirtschaftsformen des Feldbaus und der Viehwirtschaft. Ihnen gemeinsam ist eine ausgeprägte Form der Territorialität mit zentralen Orten, die sich aus den Notwendigkeiten des ortsfesten Ackerbaus (sog. Primärproduktion) und Versorgungsbedürfnissen nomadisierender Viehzüchter ergeben. Perspektivisch führte die Erfindung des Landbaus zu einer Biodiversitätslenkung und Biodiversitätsverdrängung und der Herausbildung einer anthropogenen Biodiversität (Wildformen zu Zuchtformen).

Neben den landwirtschaftlich genutzten Flächen sind Städte ökosystemare Strukturen, die sich als anthropogene Systeme nicht selbst unterhalten können. (WBGU 2000, S. 209; konkretes Fallbeispiel beispielhaft: Berger und Ehrendorfer 2011; Brunner und Schneider 2005). Städte erwiesen sich als ökologisch

folgenschwere Strukturen. Sie parzellieren das Umland nach ihren Versorgungsbedürfnissen. In europäischen Städten wurden die Grundlagen der modernen
Globalisierung (de Vries und van der Woude 1997) und der Industriellen Revolution gelegt, und damit des heutigen auch ökologischen Zustands des Weltsystems.
Die Kolonienbildung als Begleiterscheinung der Globalisierung ist ökologisch als
Ausgreifen auf die Ressourcen exterritorialer Ökosysteme zu verstehen (Sieferle
et al. 2006).

Aus Agrikultur, Urbanismus und Industrialisierung als dominierenden ökologische Strategien ergeben sich prekäre Folgen eines **Agro-Urban-Industriellen
Komplexes.** Über die strukturellen historischen Voraussetzungen des heutigen
Umweltzustands informieren die Erträge der Historischen Humanökologie (Herrmann 2016). Die Problemfelder werden hier nur als systematische Elemente
aufgeführt. Sie werden wegen der mit ihnen gegenwärtig verbundenen Herausforderungen und ethisch-politischen Konflikte im Kap. 4 aufgegriffen. Die
wichtigsten Problemfelder im Einzelnen:

- Als Folge menschlicher Handlungen und ihrer Antriebe bedroht die **gegen-
  wärtige Biodiversitätskrise** mit ihren Zuständen in terrestrischen wie marinen
  Ökosystemen perspektivisch jede Kultur (IPBES 2019; FAO 2007; FAO 2010;
  Drenckhahn et al. 2020).
- Eine der Biodiversitätskrise vergleichbare Bedrohung für ökologische Weltre-
  gimes ist der **Klimawandel** durch Treibhausgase (IPCC 2014) als Folge von
  Wirtschaftswachstum und Bevölkerungsanstieg (siehe Übereinkommen von
  Paris COP 21; Übereinkommen von Kattowitz COP 24).
- Eine ökologische Konsequenz zentraler Orte ist die Notwendigkeit, anfal-
  lende organische **Abfälle** und Fäkalien zu entsorgen, ebenso den **Müll.** Er
  ist ein Kennzeichen der manufakturellen und industriellen Produktion, dessen
  Entsorgung hochproblematisch geworden ist.
- Die anthropogenen **Massenbewegungen,** als Ausbeutung von Lagerstätten und
  Landschaftsumgestaltungen, übertreffen weltweit gegenwärtig die Massenbewegungen aus natürlichen Erosionsabläufen.
- Die bedenkenlose Freisetzung lebensabträglicher anorganischer und organisch-
  chemischer Verbindungen bedroht die Ökosysteme. Als besonders gefährlich
  gelten die Persistenten Organischen Schadstoffe (**Persistent Organic Pollu-
  tants,** POPs; https://www.pops.int/), weil ihre Verteilung und ihr Verbleib
  in der Umwelt nicht zuverlässig vorhersagbar sind und ihre Anreicherungen
  über die Nahrungsnetze bei chronischer Exposition zu Schadwirkungen im
  Endverbraucher bzw. den Konsumenten höchster Ordnung führen können.

Soziologisch-humanökologische Grundlagen 3

3.1 Ursprünge: Historische Splitter

3.1.1 Früher Beginn mit Austern: von Moebius zur Blidselbucht

„Die Auster und die Austernwirthschaft" heißt eine Studie des Kieler Zoologie-Professors Karl Moebius (1877), der darin die Überfischung der Austernbänke im Wattenmeer Schleswig-Holsteins beschreibt. Er korreliert die „Zunahme der Austernesser und Austernpreise" mit der „Abnahme der Austern": „Nichts anderes als schonungslose Befischung hat die Austernbänke entvölkert … Man wird aber fragen: Warum wurden *vor unseren Zeiten* die europäischen Austernbänke nicht übermäßig befischt? Weil es vor dem Zeitalter der Dampfschiffe, Dampfwagen und Eisenbahnen viel weniger Austernesser gab … Sobald die Austern … schnell verbreitet und tief in das Binnenland geführt werden konnten, … wuchs trotz der schnellen Steigerung der Preise der Bedarf …" (S. 88–91). Moebius kreierte dabei den Begriff der „Biocönose" (von bíos, das Leben, und koinóein, etwas gemeinschaftlich haben) oder „Lebensgemeinde" (S. 76). (Glaeser 2004; Reise 1980).

Vor allem aber stellte Moebius Regeln zur *nachhaltigen Bewirtschaftung* auf, in seinen Worten „Regeln für die Erhaltung und Verbesserung der natürlichen Austernbänke, als Grundlage aller Austernwirthschaft" (S. 124). Seine Politikempfehlung lautet: „Wenn die Austernbänke … *dauernd ertragfähig* bleiben sollen, so darf das jährliche Maass ihrer Befischung nicht nach… der Höhe der Austernpreise bestimmt werden, sondern einzig und allein nach der *Menge des Zuwachses*. Die Erhaltung der Austernbänke gehört ebenso zu den Aufgaben des Staates, wie die Erhaltung der Waldungen" (S. 125, Hervorheb. BG).

© Der/die Autor(en), exklusiv lizenziert durch Springer Fachmedien Wiesbaden GmbH, ein Teil von Springer Nature 2021
B. Herrmann et al., *Humanökologie*, essentials, https://doi.org/10.1007/978-3-658-32983-9_3

Moebius' Ausführungen sind ein frühes Lehrstück zu Humanökologie und Ressourcenökonomie, das ökologische, soziale und ökonomische Nachhaltigkeit verknüpft. Die ressourcen-orientierte Thematik ist seit den 1970er Jahren zentral für die Humanökologie, insbesondere in ihrer Göteborger Ausprägung (Kap. 3.2.3). Historisch ist anzumerken, dass die Austernzucht in der Sylter Blidselbucht, im sauberen Wattenmeer zu Dänemark, in den 1990er Jahren erfolgreich reaktiviert wurde: ein Stück planerisch umgesetzter Humanökologie, auch wenn die „neue" Auster nicht aus Europa, sondern aus dem Pazifik stammt (Reise 1998, S. 168 f.; Auskunft 22.02.2001).

3.1.2 Das Chicago der 1920er Jahre: Namensgebung in der Urbanität

Die humanökologische Geschichtsschreibung (Bruckmeier 2004) beginnt meist mit Robert Park (Park 1952; Park et al. 1925) in den 1920er Jahren. Die frühen Stadtsoziologen Chicagos, neben Park insbesondere Ernest Burgess und Roderick McKenzie, haben *Human Ecology* als interdisziplinäres Fachgebiet entwickelt. Unter Humanökologie verstand man die Untersuchung der räumlichen und zeitlichen Beziehungen menschlicher Lebewesen, bewirkt durch die selektiven, distributiven und akkommodativen Kräfte der Umwelt.

Untersuchungsgegenstand war die Stadt. Wichtige Vertreter des Gebietes waren später Otis Duncan mit Beiträgen zur Theorie der biologischen und sozialen Evolution sowie Amos Hawley (1986) zum Verhältnis von Ökologie und Humanökologie. Hawley räumte mit Biologismen und Raumanalysen in der Sozialwissenschaft auf und stellte das *specificum humanum*, menschliches Verhalten im kulturellen Anpassungsprozess, in den Vordergrund. Auch Roderick McKenzie (1968) hat in seinen späteren Arbeiten die Besonderheiten der Humanökologie mit ihrer grundlegend anderen Qualität in Abgrenzung zur Tier- und Pflanzenökologie betont (Serbser 2004).

Entscheidend ist die soziale und kulturelle Dimension menschlicher Interaktionen. Damit erfasst Humanökologie die kulturell gestalteten Bereiche *(cultural superstructure)* und die der ökologischen Ordnung *(biotic substructure)* in ihren Beziehungen. Sie bindet sozioökonomische, politische, kulturelle und naturräumliche Aspekte ganzheitlich zusammen (Teherani-Kroenner und Glaeser 2020).

3.1.3 Neubeginn auf dem Lande: das Paradigma der 1970er Jahre

In den 1970er Jahren entstand die *New Human Ecology* im Zuge der Diskussion über die ökologischen Grenzen des Wachstums und die Umweltbewegungen in den Industriekulturen des „globalen Nord-Westens", diesmal eingeleitet von Land- und Agrarsoziologen wie William Catton, Riley Dunlap und Frederick Buttel. Sie zeigte aber auch marxistische Einflüsse (Schnaiberg 1980). Das *„human exemptionalism paradigm"* (HEP) wird abgelöst vom *„new ecological paradigm"* (NEP) (Dunlap und Catton 1979, S. 250). Das neue Paradigma besagt, dass der Mensch aufhört, einzigartig *(exemptionalist)* zu sein, eine kulturelle Ausnahme-Spezies, die im Mittelpunkt des Universums oder zumindest der irdischen Biosphäre steht (Glaeser 1996a, b). Stattdessen ist er Teil einer Lebensgemeinschaft, deren Energievorräten endlich sind und die mit begrenzten globalen Ressourcen haushalten muss.

Eine umfassende Darstellung der humanökologisch orientierten Umweltsoziologie findet sich bei Riley Dunlap und William Catton (1979), weiter präzisiert bei Frederick Buttel (1987). Später forderte Catton (1993), die Soziologie insgesamt als Teil der ökologischen Wissenschaft dieser unterzuordnen, was aber keine Resonanz fand. Für die Soziologie wird eine geradezu kopernikanische Wende des Menschenbildes eingeleitet. Der Mensch rückt aus dem Zentrum der Natur und wird an deren Peripherie, als eine unter vielen *Species,* abgelegt. Begründet wird die neue Subdisziplin Umweltsoziologie, deren erster Weltkongress 2001 in Kyoto (Japan) stattfand, und die zur Konzeptbildung der Humanökologie erheblich beitrug. Die *New Human Ecology* verband die Tendenz zur biozentrischen Weltsicht mit einer politisch-normativen Komponente.

3.2   Umwelt und Globalisierung

3.2.1   Humanökologie und verwandte Disziplinen

Im Laufe der Jahre hat die Humanökologie sich aufgefächert. Verwandte und spezialisierte Paralleldisziplinen sind entstanden.

Die europäische Humanökologie, in den 1970er Jahren in Wien entstanden (Kap. 3.2.3), ist ein Abkömmling der amerikanischen Human Ecology. Im deutschsprachigen Raum gab es zunächst Berührungsängste, eine Nähe zur Biologie sollte vermieden werden. Der Soziologe René König (1958, S. 264 f.) machte die Human Ecology, die er aber in „Sozialökologie" umtaufte,[1] bekannt. Damit hat er zwar den gesellschaftlichen Bezug verdeutlicht, aber für eine Zweiteilung der Disziplin gesorgt. Seither existieren im deutschen Sprachraum beide nebeneinander, in Wien die Humanökologie der Schule Helmut Knötig mit biologischer Herkunft, und die dem Gedanken des Metabolismus verpflichtete soziologisch orientierte Sozialökologie von Marina Fischer-Kowalski. In Frankfurt trat 1989 die von Egon Becker initiierte soziale Ökologie des ISOE-Instituts hinzu, mit dem Schwerpunkt gesellschaftlicher Naturverhältnisse.

Kulturökologie bezieht den Menschen als Kulturwesen ein und betrachtet die Mensch-Natur-Beziehungen als Folge kultureller Leistungen (Steward 1955). Natur wird thematisiert, doch nicht die vom Menschen unberührte, die ‚intakte' Natur, sondern die gestaltete ‚kulturierte' Natur. Das Thema ist Einheit von Natur und Kultur, und zwar in der Kultur (Glaeser 1992, S. 61).

Ethnoökologie, aus der Tradition der sozialen Anthropologie und Ethnologie entstanden, befasst sich zunächst mit bäuerlichen und nomadischen Gemeinschaften (Rappaport 1968). Vielfach geht es um die Anerkennung lokalen Wissens *(indigenous knowledge systems)*, verbunden mit Fragen zum Schutz der biologischen Vielfalt und der Biodiversität (Teherani-Krönner und Glaeser 2020, S. 411 f.).

Politische Ökologie untersucht die Auswirkungen menschlichen Handelns auf Ökosysteme, die Wechselwirkungen zwischen abiotischen, biotischen und gesellschaftlichen Faktoren mit Bezug auf den jeweiligen politischen Rahmen (Egger und Glaeser 1975).

3.2.2 Das Bedürfnis nach Bindestrich-Ökologien mit normativer Tendenz

Die Bedürfnis nach „Bindestrich-Ökologien" scheint vom Wunsch nach Klarheit und theoretischer Eindeutigkeit getrieben. Der Grund ist eine gewisse Unschärfe, eine Mehrdeutigkeit divergenter Funktionen des Begriffs „Ökologie". Konkret

---

[1]König, Herausgeber des Handbuchs der empirischen Sozialforschung, bevorzugte „Sozialökologie" (mündliche Mitteilung von Rainer Mackensen 1989 an Parto Teherani-Krönner; https://link.springer.com/chapter/10.1007%2F978-3-663-14614-8_3 (aufgerufen am 1.11.2019).

versteht sich der Ökologiebegriff bei Ernst Haeckel (1866, S. 286)[2] als holistische Funktionalität, als Beziehungsgeflecht aller Organismen mit ihrer Umwelt, woraus Anpassungen resultieren. Das wissenschaftlich-theoretische Erfassen der Erfahrungsobjekte enthält eine Erkenntnistheorie, welche auf die apriorischen Bedingungen der Möglichkeit solcher Erkenntnisse abzielt. Andererseits fragt die politisch-institutionelle Anwendung nach den normativen Konsequenzen der funktionalen Geflechte für eine zu erhaltende und zu schützende Umwelt.

Dem suchte die Humanökologie Rechnung zu tragen mit ihren Spezifizierungen in Richtung auf Soziales, Kultur, Ethnizität oder Politik – mit dem Resultat der entsprechenden Bindestrich-Ökologien. Sie sind Folge des Schwankens zwischen analytisch-empirischer Beschreibung und Erklärung einerseits und normativer Wünschbarkeit unter ethisch-politischen Vorzeichen andererseits. Die frühe Chicagoer Humanökologie der 1920er Jahre befasste sich exemplarisch mit Urbanität. Erkenntnisgegenstand war eine Theorie des Beziehungsgeflechts zwischen einer Mensch-Gesellschaft-gebauten und einer natürlichen Umwelt. Die *New Human Ecology* der 1970er Jahre befasste sich exemplarisch mit Ruralität. Sie baute auf den früheren Theorien auf, versah den Umweltbegriff jedoch mit einer normativ-politischen Komponente im Einklang mit der aufkommenden Umweltbewegung, die Analysen der Grenzen weltwirtschaftlichen Wachstums eingeschlossen (Teherani-Krönner und Glaeser 2020, S. 412).

Beiden gemeinsam ist die Überzeugung einer „transzendentalen" (Immanuel Kant), vor aller Erfahrung liegenden (apriorischen) Theorie des Erkennens dieser Zusammenhänge und damit der teils verdeckte, teils offen dargelegte Anspruch einer Wissenschaft, sich von den herkömmlichen „cartesischen" Wissenschaften der unterkomplexen „Subjekt-Objekt-Beziehungen" von Grund auf zu unterscheiden. Die Idee einer rationalen, aber neuartigen Wissenschaftswissenschaft, die über Theoriebildung und empirisch-konkrete Anwendungen hinaus, auch in Fallstudien, normativ in Sollensforderungen und Politikbereiche mündet, faszinierte viele Humanökologen.

---

[2]„Unter Oecologie verstehen wir die gesammte Wissenschaft von den Beziehungen des Organismus zur umgebenden Aussenwelt, wohin wir im weiteren Sinne alle ‚Existenz-Bedingungen' rechnen können. Diese sind theils organischer, theils anorganischer Natur; sowohl diese als jene sind, wie wir vorher gezeigt haben, von der grössten Bedeutung für die Form der Organismen, weil sie dieselbe zwingen, sich ihnen anzupassen."

3.2.3 Aufbruch zur interdisziplinären Umweltforschung und Schulbildung

In den späten 1960er Jahren begannen sich die Umweltbewegungen in USA und Nordwest-Europa zu etablieren. Der *Club of Rome* stellte die Frage nach den Grenzen wirtschaftlichen Wachstums und finanzierte mithilfe der Stiftung Volkswagenwerk die erste globale Studie zu diesem Thema (Meadows et al. 1972). Die erste Welt-Umwelt-Konferenz, die *„United Nations Conference on the Human Environment"*, ereignete sich 1972 in Stockholm und machte die widerstreitenden globalen Nord-Süd-Interessen zum Thema: Umwelt hier, Entwicklung dort.

Die Soziologie tat sich mit Natur und Umwelt seit jeher schwer, da sich – so Emile Durkheim (1858–1917), einer der Begründer des Faches – soziale Fakten nur durch andere soziale Fakten erklären lassen. Humanökologie etablierte sich akademisch, zeitgleich und zum Teil inhaltlich überlappend, mit der Umweltsoziologie, in den 1970er Jahren durch die Gründung nationaler und internationaler wissenschaftlich-multidisziplinärer Gesellschaften. Angestoßen wurde diese Entwicklung durch die internationalen Konferenzen und Workshops, die zu Beginn der 1970er Jahre von der *„International Organisation for Human Ecology"* (IOHE) an der TU Wien durchgeführt wurden. Ihr *spiritus rector* Helmut Knötig (1976) begründete die „Wiener Schule" der Humanökologie mit systemökologischer Ausrichtung und einem umfassenden Wissenschaftsanspruch. Die Teilnehmer kamen aus allen Kontinenten, auch aus Osteuropa und der Sowjetunion, was in Zeiten des „kalten Krieges" an kaum einem anderen Ort der Welt politisch möglich gewesen wäre (Glaeser 2004, S. 29–31).

Humanökologisch forschende Wissenschaftler verschiedener Disziplinen erörtern auf den jährlichen Tagungen der Deutschen Gesellschaft für Humanökologie (DGH) (https://www.dg-humanoekologie.de) aktuelle Themen der Mensch-Umwelt-Beziehung. Die Berichte und Ergebnisse erscheinen in der Fachzeitschrift GAIA und in Sammelbänden beim Oekom-Verlag. Humanökologische Zentren und universitäre Einrichtungen entstanden in vielen Ländern rund um den Globus. In China verband das *„Department of Systems Ecology"* der *Academia Sinica* unter dem verstorbenen Leiter Wang Rusong altchinesische Traditionen mit mathematischen Modellansätzen und empirisch-ökologischer Sozialforschung in urbanen und ruralen Regionalstudien. Dauerhaft erfolgreich war die *„Society for Human Ecology"* (SHE) mit Basis in den USA und regelmäßigen Tagungen in verschiedenen Ländern (Suzuki et al. 1991; Borden und Jacobs 1989).

In der Folge wurde das Fach Humanökologie in den Universitätsbereich eingeführt. Schweden richtete vier Lehrstühle ein mit unterschiedlichen Schwerpunkten in Lund, Göteborg, Uppsala und Umeå. Die Universität Göteborg siedelte den humanökologischen Studiengang in einer Fakultät für „thematische Studien" an, die wiederum von den drei klassischen Fakultäten für Natur-, Geistes- und Sozialwissenschaften betreut wurde. In den Vereinigten Staaten wurde mit dem College of the Atlantic in Bar Harbor (Maine) ein Liberal Arts College geschaffen, das insgesamt der Humanökologie gewidmet ist. Bis auf Bar Harbor wurden die meisten dieser akademischen Einrichtungen geschlossen oder anderswo angesiedelt (Bruckmeier 2004, S. 77 f., 83–85, 98–100).

In Deutschland wurden kaum humanökologischen Professuren eingerichtet. Humanökologische Forschung wurde im Namen anderer Disziplinen (Geographie, Biologie, Anthropologie, Medizin oder Geschichtswissenschaft) sowie in außeruniversitären Forschungseinrichtungen (WZB) betrieben (Teherani-Krönner und Glaeser 2020, S. 411).

Zusammenfassend erschließt sich die Humanökologie als Theorie des Wechselspiels zwischen Gesellschaft und Umwelt, Mensch und Natur in folgenden Bezugslinien und Horizonten:

- im interdisziplinären Bezug, der auch die Methodenfrage neu stellt (Kap. 3.2.4);
- im internationalen Austausch, Industrie- und Entwicklungsländer eingeschlossen;
- durch problemorientierte Forschungsfragen, die einerseits politikrelevant sind, andererseits die Betroffenen aus der lebensweltlichen Praxis transdisziplinär einbeziehen;
- nicht zuletzt schließlich in der kritischen Reflexivität ihres Vorgehens.

3.2.4 Methoden interdisziplinär-humanökologischer Forschung

Methodenbildung ist ein disziplinäres Kennzeichen, zur Interdisziplinarität gehöre dagegen kein spezifisches Methodengefüge. Diese Aussage ist nur bedingt zutreffend, da mit zunehmend interdisziplinär orientierten Forschungsarbeiten auch die Erfahrung zunimmt, wie derartige Kooperationen methodisch handhabbar werden. Im Bereich der Humanökologie gab es hierzu immer wieder explizite Vorstellungen (Glaeser 2006).

Ein früher Versuch der Systematisierung methodologischen Wissens und Vorgehens zur Durchführung interdisziplinärer Studien findet sich bei Richard Brislin (1980). Umweltforschung ist grundsätzlich problemorientiert und interdisziplinär. Wichtig für die Zusammenarbeit ist es, interdisziplinäre Hypothesen zu bilden, welche die Methodenwahl leiten, nicht umgekehrt. Mit dem Überschreiten der Disziplingrenzen steigt die Zahl unabhängiger qualitativer wie quantitativer Variablen (Glaeser 2004, S. 33–36).

Ab 1996 wurde im Anschluss an die erste Berufung auf den humanökologischen Lehrstuhl der Universität Göteborg (Bernhard Glaeser) dem Aspekt Methodik gezielte Aufmerksamkeit geschenkt, der sich in Forschungsprogramm und Ausbildung der Studierenden niederschlug. Der Ansatz wurde für die innovative schwedische Küstenforschung, insbesondere in der Kooperation zwischen Natur- und Sozialwissenschaftlern, in dem Programm SUCOZOMA nutzbar gemacht (Glaeser 1999; Kap. 3.3). In der deutschen Küstenforschung konnten Modelle und Szenarien zum gemeinsamen Verständnis über biologische, historische und soziale Bezüge in Gesellschaft-Natur-Systemen beitragen (Glaeser et al. 2007, S. 299 f.).

Stakeholder-Beteiligungen[3] sind ergänzend unter methodischen Gesichtspunkten zu beachten. Bei Beobachtungen oder in Interviews sind Forscher auf Administratoren, Lobbyisten, Anwohner und andere Beteiligte angewiesen. Das Vorgehen ist im „Projektgebiet" bekannt zu geben. Dem dient eine standardisierte mündliche oder schriftliche Einführung, die Forschungsziele, Nutzen für die Betroffenen und „Beforschten" erläutert (Glaser et al. 2010).

3.2.5   Kritische Theorie: ein Neuansatz?

Trotz starker Verwurzelung der Humanökologie in der Soziologie hat die „kritische Schule" der Soziologie hier kaum Fuß gefasst. Als Ausnahmen gelten die politisch-ökonomische Kritik Allan Schnaibergs (1980) an den ökologisch verheerenden Konsequenzen des Kapitalismus oder die feministisch-ökologische Revolution in der Wissenschaftstheorie bei Carolyn Merchant (1980). Umso mehr verblüfft der wenig beachtete kritisch-methodische Ansatz von Raymond Morrow und David Brown. Letztere bemängeln, dass der kritische Wissenschaftsmodus eine Fülle kreativen Denkens hervorgebracht habe, doch wenig empirische Forschung. Für sie ist die entscheidende Frage: Wie lässt sich kritische Theorie in Forschungspraxis umsetzen und wie unterscheidet sich kritische Methodologie von „empirizistischen" Ansätzen (Morrow und Brown 1994, S. 268)?

---

[3] Als Stakeholder gelten Personen, Personengruppen oder Institutionen bis hin zu Regierungen, die einen Prozess beeinflussen können oder von ihm beeinflusst werden.

Das Spezifikum kritischer Forschung liege in seiner nicht-empirischen Reflexivität, kombiniert mit Fallstudien-Methodik, zum Zwecke vergleichender Verallgemeinerung, Interpretation und Erklärung. Kritische Forschung, auch im interdisziplinären Umweltbereich, folge dem Anspruch, dass soziale Realität – etwa als Handlungen, Prozesse, Institutionen – der Legitimation bedarf, genau so und nicht anders zu sein. Der methodische Aspekt wird durch den Begriff der *careful explication* eingeführt; er hat geringe Folgen gezeigt (Glaeser 2004, S. 37 f.).

3.3 Von den Konzepten zur empirischen Forschung: Küsten mit Humanökologen nachhaltig entwickeln

Küsten entstehen dort, wo Erde, Wasser und Luft aufeinander treffen. „Küste" hat einen Doppelcharakter: Sie ist eine gegenständliche, physische Größe und zugleich mentales Konstrukt. Ihre kulturhistorische Bedeutung, ihre biologische und wirtschaftliche Produktivität gehen einher mit ökologischer Verletzlichkeit. Vielfältige Nutzungen sind mit unterschiedlichen Interessen verknüpft und schaffen Konflikträume. Hier setzt das Integrierte Küstenzonen-Management (IKZM) ein, um zu vermitteln und nachhaltige Entwicklung zu ermöglichen. Nachhaltigkeit ist eine „regulative Idee" (sensu Immanuel Kant), deren normsetzende Kraft ethisch zu begründen ist. Dieser Zielsetzung ist auch der humanökologische Ansatz verpflichtet, der den Bogen schlägt von der reflexiven Begründung bis zur Umsetzung von Nachhaltigkeit im Küstenmanagement (Glaeser 2005, S. 9).

3.3.1 Küsten im historischen Kontext: Warum IKZM?

Seit einigen tausend Jahren spielen Küsten eine bedeutende Rolle im Zuge der Menschheitsentwicklung: als Zentren des Handels, bei Städtegründungen, als politische Machtzentren und für die Entstehung von Hochkulturen (Paine 2013). Die europäische Philosophie entstand mit der vorsokratischen Naturphilosophie von Thales, Anaximandros und Anaximenes in Milet, an der Küste Kleinasiens. Thales (624–545 v. u. Z.) erklärte die Weltentstehung aus Wasser. Seine Begründung ist nicht überliefert. Denkbar ist, dass das Meer als Voraussetzung für Seehandel und Reichtum seiner Heimat – der Urvater hieß Okéanos – ihm die Wahl des Wassers nahe legte (Scholtz 2016, S. 13–25).

Das Meer erregte seit der Antike Abscheu und Angstgefühle als Ort des Chaos mit der ‚Erinnerung' an die Sintflut. Erst im 17. Jahrhundert setzte eine Umdeutung ein. Sehnsucht nach Meeresküsten entstand im Europa des 18. und 19. Jahrhundert, „Meereslust" in Literatur, Malerei, Wissenschaft, Medizin und nicht zuletzt im gesellschaftlichen Leben (Corbin 1994). Neue Schrecken für Küsten verbreitet, bedingt durch die Erwärmung und Ausdehnung der Meere, der Klimawandel, der auch kulturelle und soziale Herausforderungen mit sich bringt (Fischer und Reise 2011). Das „Anthropozän", das menschbeherrschte Erdzeitalter, gilt Vielen als das nahende Ende der globalen Bio- und Soziosphäre. Der Weltklimarat IPCC (https://www.de-ipcc.de/) generiert seit Jahrzehnten Daten zum Verstehen und Eindämmen des Klimawandels. Klaus Meyer-Abich (1972) warnte bereits in den 1970er Jahren vor einem globalen Temperaturanstig von >1 % C. In einer beispiellosen Lobby-Kampagne gelang es US-Wissenschaftlern 1989, die Weltpolitik von den Gefahren des Klimawandels zu überzeugen. In Nordwijk (November 1989) sollte der *„Carbon Accord"* unterzeichnet werden, was an der mangelnden Bereitschaft der USA scheiterte (Rich 2018).

Heute ist auf internationaler Ebene die Politik der Küsten und Meere Bestandteil nationaler Interessen und Thema zwischenstaatlicher Vereinbarungen. IKZM ist Teil nationaler und regionaler Entwicklung und Forschung. Vorreiter der Küstenforschung in Europa war das nationale schwedische SUCOZOMA (SUstainable COastal ZOne MAnagement) 1997–2003 auf humanökologischer Basis. Nach dem schwedischen Modell waren die BMBF-finanzierten deutschen Verbundprojekte „Nordsee" *(Zukunft Küste – Coastal Futures)* und „Ostsee" *(IKZM Oder)* konzipiert (beide 2004–2010: Kannen et al. 2010), die wiederum die Grundlage bildeten für die nationale Strategie für Küsten und Meere für Deutschland (Glaeser et al. 2004).

3.3.2 Der Fall SUCOZOMA: Nachhaltige Küstenentwicklung in Schweden

Das schwedische Forschungsprogramm SUCOZOMA untersuchte Probleme der Ressourcennutzung und des Umweltschutzes in schwedischen Küstengebieten. Die Nutzungsprobleme betrafen sowohl lebende (Fischerei und Muschelzucht) als auch nicht-lebende Naturressourcen (Wasser und Boden) (Bruckmeier 2005, S. 56). Die Feder führte das Humanökologische Institut der Universität Göteborg, in der Initialphase unter der Leitung von Bernhard Glaeser. Beteiligt

waren unterschiedliche Disziplinen wie 1) Humanökologie, ökologische Ökonomie, Politikwissenschaft; 2) Marinebiologie, Toxikologie, Ökologie; 3) Fischerei-
und Meeresbiologie, Genetik (Abb. 3.1).

Die Zusammenarbeit zwischen Natur- und Sozialwissenschaften in SUCO-
ZOMA war durch das Konzept der strategischen Umweltforschung und die
vermittelnde Rolle der Göteborger Humanökologie geprägt. Andererseits wurden
nicht-wissenschaftliche Nutzer bei Problemformulierungen und der Bewertung
von Nutzungskonflikten hinzugezogen. Hier war die SUCOZOMA-Forschung
nicht nur inter-, sondern auch transdisziplinär prägend (Bruckmeier 2005,
S. 77 f.).

3.3.3   Küsten zwischen Ethos und Management

Jede Optimierungsentscheidung zwischen den drei Nachhaltigkeitspfeilern Ökologie, Ökonomie und Soziales stellt eine Güterabwägung dar, die letztlich ethisch
zu begründen ist: Welche Küste sollen wir wollen? Ein Beispiel für eine Ethik
der Nachhaltigkeit ist die *environmental virtue ethics* dar, eine umweltbezogene
Tugendlehre von Philip Cafaro (2001). Wertorientierte Umweltethik stellt sich
den Fragen eines (normativ) richtigen Umgangs mit Natur. Im Unterschied zur
Wertewahrheit, die ethische Wahrheit durch intrinsische Werte begründet, bevor-
zugen Diskursethiker eine **argumentative Wahrheit,** welche durch sinnvolles
Argumentieren herausgefiltert wird (Glaeser 2005, S. 16, 21).

Im Küstenkontext stellt sich die Frage: Wie ermittelt der ethische Entwurf
konkret anwendbare Normen? Hierzu hält das amerikanische National Research
Council (1995, S. 7–12) zum Verhältnis von *„coastal science and policy"* fest,
dass Forschung idealerweise zwei Funktionen hat:

- natürliche und menschliche Systeme zu verstehen und zu erklären;
- deren Interaktionszusammenhang auf sozial erstrebenswerte Weise zu struktu-
  rieren.

Hier gehen also neben dem grundlegenden Systemverständnis normative Vorgaben ein, die ohne ethische Wertsetzungen nicht ableitbar sind. An diesen
Prozessen sind deutlich unterschiedene „Kulturen" beteiligt. Insoweit lässt sich
von einer Kulturökologie des *public policy making* an der Küste sprechen (Glaeser
2005, S. 21).

Program structure

Integrated management

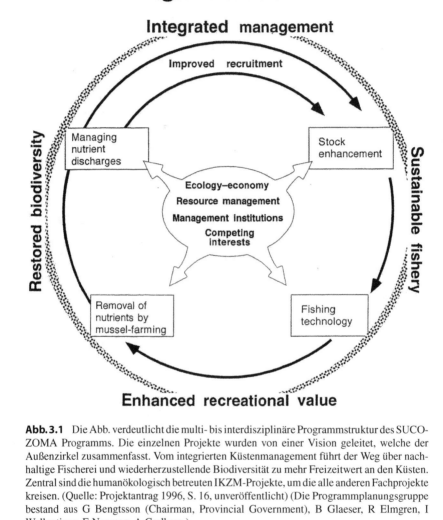

Abb. 3.1   Die Abb. verdeutlicht die multi- bis interdisziplinäre Programmstruktur des SUCO-ZOMA Programms. Die einzelnen Projekte wurden von einer Vision geleitet, welche der Außenzirkel zusammenfasst. Vom integrierten Küstenmanagement führt der Weg über nachhaltige Fischerei und wiederherzustellende Biodiversität zu mehr Freizeitwert an den Küsten. Zentral sind die humanökologisch betreuten IKZM-Projekte, um die alle anderen Fachprojekte kreisen. (Quelle: Projektantrag 1996, S. 16, unveröffentlicht) (Die Programmplanungsgruppe bestand aus G Bengtsson (Chairman, Provincial Government), B Glaeser, R Elmgren, I Wallentinus, E Neuman, A Carlberg.)

3.4   Ausblick: Das humanökologische Paradigma

Die Humanökologie vertritt einen (wissenschaftsimmanent) andersartigen „Denkstil" (Karl Mannheim), der sich auszeichnet durch ganzheitliches statt reduktionistisches, organisches statt mechanisches Denken, Wechselwirkung statt Kausalität. Die Zweckmäßigkeit der Lebensprozesse im Hinblick auf ein *hólon,* ein Ganzes, ist das erkenntnistheoretische Modell. In den 1920er Jahren war Humanökologie eine Vorreiterin bei der Analyse sozialen Elends im korrupten Chicago, immer im Kontext der Dreiecksbeziehung „Mensch-Gesellschaft-Umwelt", **ohne Wertvorgaben.** Die Umweltkomponente der Humanökologie in den 1970er Jahren konnte, jetzt **umweltpolitisch bewertend,** auch gesellschaftlich ‚nutzbar' werden (Teherani-Krönner und Glaeser 2020, S. 412). Auf diesem Nährboden entstanden neuere Ansätze mit ähnlichem Anspruch, etwa *sustainability science, ecological economics, social-ecological systems analysis* oder *earth system analysis* (Glaser et al. 2012, S. 199).

Der nachdrückliche Erfolg der Humanökologie als Ideengeberin ging einher mit einer institutionellen Krise in den 1990er Jahren.[4] Während Umwelt, Ökologie, ganzheitliches Denken *mainstream* wurden, wissenschaftlich wie gesellschaftlich, verschwanden humanökologische Institute an Universitäten (Bruckmeier 2004, S. 77 f., 92 f.). Eine Erklärung hierfür steht noch aus. Aus der Krise könnte jedoch eine Erneuerung erwachsen, eine kritische selbstreflexive und interdisziplinäre Wissenschaft, welche Zukunftsthemen wie Digitalisierung oder Gentechnik aufgreift. Die DGH hat bereits damit begonnen.

---

[4]Bemerkenswert ist hieran, dass der Philosophie Ähnliches nachgesagt wird. Als ursprünglich einzige Wissenschaft neben den Universitätsdisziplinen Theologie, Rechtswissenschaft und Medizin schuf sie die Grundlagen für spätere Spezialwissenschaften, die sie in Form von ‚Ausgründungen' aus sich entließ. Das begann mit den Naturwissenschaften und setzte sich fort mit historischen, Sozial- und Kulturwissenschaften.

Ethisch-humanökologische Grundlagen   4

4.1   Warum Ethik und wie?

Nicht erst seit Beginn des 21. Jahrhunderts sind die massiven Verluste von
Arten sowie zerstörerische Veränderungen und extreme Flächenrückgange prak-
tisch aller Ökosystemtypen aufgrund des Umfangs und der Art von Wirkungen
des *Agro-Urban-Industriellen Komplexes* (vgl. Kap. 2.5) gut dokumentiert. Dies
ist für eine Humanökologie, die „Wechselbeziehungen zwischen den biologischen
Grundbedürfnissen und kulturell moderierten Lebensansprüchen der Art *Homo
sapiens*" sowie „den von Menschen genutzten Ökosystemen bzw. der Biosphäre"
(vgl. Kap. 5) im Blick hat, von höchster Relevanz. Zum ersten bildet es heute
vielleicht sogar den zentralen Gegenstand der Forschung selbst, zum zweiten
stellt sich die Frage, was dies für aktuelles und künftiges menschliches Handeln
bedeuten kann – und soll.

Es ist zu erwarten, dass weiter die mittel- und langfristigen Lebens- und Über-
lebenschancen für viele Spezies im Kontext ihrer Ökosysteme deutlich geringer
werden. ‚Daher' – so heißt es oft – sind sofortige Maßnahmen zur Sicherung
der Ökosysteme erforderlich für eine (auch) künftig noch lebenswerte Welt. Um
dies allerdings zu einem gültigen Argument zu machen, reichen natur- und sozialwissenschaftliche Analysen und Argumente nicht aus: Eine normative oder
Sollensaussage – also, dass etwas ge- oder verboten, erlaubt oder (nicht) wün-
schenswert ist – benötigt einen normativen Grund. So könnte mit Hans Jonas
ausgeführt werden, warum es geboten ist „dass eine Menschheit sei" (Jonas
1979, S. 91). Dann wäre die moralische Verpflichtung zur Selbsterhaltung der
Menschheit ein Grund für spezifische Normen und Maßnahmen, die wiederum
unter interdisziplinärer Mitberücksichtigung natur- und sozialwissenschaftlichen

Wissens genauer zu bestimmen sind. Ethik als philosophische Reflexions- und Begründungstheorie der Moral kommt hier ebenso ins Spiel wie das Recht, das staatlich kodifizierte Regelwerke interpretiert und kontrolliert, sowie die Politik mit Parlamenten als Normsetzungs- und Verwaltung als Umsetzungsinstanz.

Zur Begriffsklärung: *Moral* bezeichnet individuelle und/oder kollektive Werte und Normen von Menschen, die Vorstellungen über das Gerechte und das Gute umfassen. Moralisch richtig ist, was zumindest *auch* um seiner selbst willen getan werden soll, nicht *nur* zu einem externen Zweck. *Ethik* ist eine philosophisch-interdisziplinäre Disziplin. Sie analysiert die Moral als ‚gelebte Sittlichkeit‘, aber sie sucht darüber hinaus nach plausiblen Begründungen für moralische Werte und Normen sowie konkrete moralische Urteile. Wie jede Wissenschaft kann Ethik keine Letztbegründungen liefern, aber kritisch prüfen, welche Position wie weit und auf Basis welcher Vorannahmen argumentativ plausibel und nachvollziehbar, bzw. plausibler als andere ist. Für die Humanökologie sollte die anwendungsbezogene Ethik ein notwendiger interdisziplinärer Bestandteil sein (vgl. Bruckmeier und Serbser 2008), selbst wenn bzw. gerade weil ihr Einbezug länger ein Desiderat blieb.

Die Frage der moralischen Bedeutsamkeit der nichtmenschlichen Dinge der Welt wird unter verschiedenen, nicht immer trennscharfen Ausdrücken gefasst: *Naturethik* ist ein übergreifender Ausdruck, der neben *grundlegenden Reflexionen des Mensch-Natur-Verhältnisses* praxisbezogen *Ressourcenethik, Tierethik,* und *Naturschutzethik* umfasst. *Umweltethik* wird teils umfassend im Sinne von Naturethik verstanden, teils aber enger im Sinne von Ressourcen- plus Naturschutzethik *und nur mit Blick auf die Umwelt des Menschen.* Der Ausdruck *ökologische Ethik* ist verbreitet, jedoch problematisch, weil Ökologie eine *empirische* Wissenschaft ist, die nicht zugleich ein Modus der Ethik sein kann (Eser und Potthast 1999 und Vorwort). Ebenfalls normative, aber eher rechtliche und politische Fragen werden unter dem Begriff *Politische Ökologie* verhandelt, mit all den damit verbundenen Ambivalenzen. Solche unterschiedlichen Benennungen sind nicht banal oder zufällig, sondern sie zeigen unterschiedliche Perspektiven und Schwerpunkte des Zugangs an.

4.2   Ethisch-politische Mensch-Natur-Verhältnisse

4.2.1   Metaphysische und naturphilosophische Grundfragen

Die Frage der moralischen Bedeutsamkeit von nicht-menschlichen Wesen und Dingen der Welt ist wohl so alt wie menschliche Kultur. Sie ist anscheinend primär in kosmologischen spirituellen und religiösen Systemen verortet, die das *Mensch-Natur-Verhältnis* insgesamt ordnen, weit über die moralische Dimension hinaus (Cassirer 1996 u. v. a.). Eine kategoriale Trennung zwischen Menschen und Natur – zwischen Geister- und Götterwelt, Menschen und anderen Naturwesen – selbst ist keinesfalls eine menschliche Grundkonstante, sondern das Produkt der westlichen Moderne (Descola 2013). Die wertbezogene strikte Trennung zwischen Menschen und Natur sowie speziell die Variante in der Kombination von christlicher Religion, patriarchalem Denken und wissenschaftlicher Rationalität werden nicht selten als mentale ‚tiefe‘ Ursachen für die heutigen Umweltkrisen gesehen (Merchant 2020 u. v. a.). Die Humanökologie muss solche Debatten aufnehmen, ist zugleich als Forschungsfeld dabei dem ‚westlichen‘ Wissenschaftsverständnis grundsätzlich, aber keinesfalls unkritisch, verpflichtet (vgl. Kap. 1). In diesem Sinne wird auch die Frage nach (Human)Ökologie und Ethik im Folgenden auf die wissenschaftlich-technische (Post)Moderne fokussiert – bei voller Anerkennung der historisch und global sehr viel reichhaltigeren Motivlagen und Quellen naturethischen Denkens.

4.2.2 Klugheitsethik: Regelwerke zur Nutzungskontrolle

Die drohende Übernutzung von ökologischen Systemen hat in der Geschichte zu moralisch-rechtlich-politischen Regelwerken geführt, um *Zugangs- und damit Ressourcenkontrolle* zu erwirken. Neben dem exklusiven – nicht selten mit Gewalt durchgesetzten – Modell Privateigentum gibt es unterschiedliche Gemeinschaftsmodelle der Nutzung. Oft wird behauptet, dass diese aufgrund des Egoismus aller Nutzenden nicht funktionieren könnten („Tragik der Allmende“). Wie jedoch die Wirtschaftsnobelpreisträgerin Elinor Ostrom (1999) zeigen konnte, ist dies nicht der Fall, solange der Allmendezugang nicht völlig regellos vollzogen wird. Das ethische Motiv hinter allen Modellen ist das *aufgeklärte Eigeninteresse:* Wenn wir (meine Gruppe, Familie, Dorf, inkl. Nachkommen) langfristig natürliche Ressourcen nutzen wollen, dann ist es klug, sich so und zu verhalten und entsprechende Regelwerke für Alle aufzustellen. Es handelt sich hier um *klugheitsethische (prudentielle) Argumente*. Dies gilt auch für die berühmt gewordene forstliche Nachhaltigkeitsidee von Hanns Carl von Carlowitz (1713), nur so viel Holz zu entnehmen, wie im entsprechenden räumlichen und zeitlichen Bezugsrahmen wieder nachwächst. Dazu gehörten allerdings weitere Schutzmaßnahmen wie Bodenerhaltung, ggf. Jagd u. a., um das Nachwachsen zu ermöglichen. Geschützt wird in all diesen Modellen allein die von Menschen genutzte Natur in Form von (stofflichen) Ressourcen. Das *ethische Klugheitsprinzip* wird weit jenseits individueller Handlungen als Grundlage für politische, letztlich rechtliche Regelwerke der Gesellschaft zugrunde gelegt, und dies bereits seit dem Altertum (vgl. Herrmann 2016).

4.2.3 Naturschutz(ethik): Reaktion auf die industrielle Moderne

Die technisch-industrielle Moderne mit entsprechenden Veränderungen der Landwirtschaft sowie die Entstehung urbaner Räume führten im 19. Jahrhundert – oft in Verbindung mit naturgeschichtlichen (z. B. Alexander von Humboldt 1769–1859) und literarisch-philosophischen (z. B. David Henry Thoreau 1817–1862) Entwürfen – zu einem neuen Typus naturethischer Argumente: Zum Ersten entstand auf Basis unterschiedlicher Ethik-Konzepte (vgl. Kap. 4.3) der *Tierschutz,* der sich gegen die Grausamkeit im Umgang mit (manchen) leidensfähigen Wirbeltierarten richtete.[1] Mit Bezug auf Vögel kamen dazu Überlegungen zur *Nützlichkeit* der ‚gefiederten Freunde' als Schädlingsvertilger. Zum Zweiten wurde der *ästhetische Wert* von Tier- und Pflanzenarten sowie die Sehnsucht nach *Wildheits-Erfahrungen* wurden zum Schutzmotiv. Nicht zuletzt jedoch ging es dem frühen Naturschutz ab Mitte des 19. Jahrhunderts zum Dritten um die *Erhaltung der heimatlichen Landschaft,* in der die mit einfachsten Mitteln wirtschaftenden Bauern durchaus der Natur zugerechnet wurden. Zugleich sollte die Landschaft nicht durch sichtbare Zeichen der Modernisierung (Werbetafeln, Seilbahnen, Staudämme) ‚verschandelt' werden. Das politische Programm war zunächst rückwärtsgewandt und hing an der Feudalgesellschaft; Städte galten nicht nur als naturfern, sondern auch als Brutstätten der Sozialdemokratie; zudem war der Schutzimpuls durchaus elitär – Naturschönheiten sollten nicht durch Massenbesuch verhunzt werden (vgl. Schmoll 1994). Zugleich ist der Naturschutz stets auch eine Angelegenheit *wissenschaftlicher Expertise* aus Geographie, Botanik und Zoologie gewesen; seit den 1920er Jahren trat die Ökologie hinzu (Potthast 2006). Der Schutz der vermeintlich ‚echten' *Wildnis* spielte in Mitteleuropa nur in kleinen Gebieten eine Rolle, anders war dies auf dem Amerikanischen Doppelkontinent, wobei allerdings die Nutzungsgeschichte der nichteuropäischen Menschen oft ignoriert wurde (Butzer 1992; Cronon 1995). Die Motive zur Ausweisung großer Wildnisgebiete waren oft der Wert als *nationales Symbol* und für

---

[1]Ins späte 18. Und 19. Jahrhundert fällt auch die ethisch-menschenrechtlich begründete Forderung nach Abschaffung der Sklaverei.

den *Tourismus* (Ästhetik, spektakuläre Wildtiere) sowie, vor allem in den russischen bzw. sowjetischen „Zapovedniks", auch die Möglichkeit der *Forschung* an möglichst allen *natürlichen Ökosystemen* und der dort ungestört ablaufenden *Evolution* (Shtilmark 2002).

4.2.4 Umweltschutz und „Grenzen des Wachstums": die Umweltkrise als ethische Herausforderung

Insbesondere mit der Aufsehen erregenden Publikation des Club of Rome zu den „Grenzen des Wachstums" (Meadows et al. 1972), die nach Erscheinen insbesondere aus Teilen der herrschenden Ökonomik ausgesprochen harsch zurückgewiesen wurde, bekam auch die Perspektive einer systematischeren Integration von Fragen des Umweltschutzes und der Wirtschaft und, darüber hinaus der gesellschaftlichen Ausrichtung von Landwirtschaft, Technik, Reproduktion und Konsum einen erheblichen Schub. Die 1970er Jahre waren auch Gründungsjahre der Humanökologie (vgl. Kap. 3.2). Die (An)Erkenntnis der globalen Bedrohung stimulierte zugleich die Philosophie, der vernachlässigten Verbindung von Umwelt und Ethik nachzugehen und ein neues Feld – *environmental ethics* – zu prägen. Neben grundlegenden Fragen des Mensch-Natur-Verhältnisses werden dabei Begründungstheorien zur moralischen Bedeutung der Natur ebenso wie konkrete Anwendungsfelder aufgearbeitet (vgl. Ott et al. 2017).

4.3 Der UN-Brundtland-Bericht und die Konferenz von Rio: Von kluger Ressourcennutzung zu Nachhaltiger Entwicklung als globaler Umweltgerechtigkeit

In den 1980er Jahren wurde deutlich, dass es in der internationalen, globalen Politik zwei Konfliktbereiche gab, die als getrennt oder gar als Antagonisten galten: Der wirtschaftlich und gesellschaftlich verstandene Entwicklungsgedanke ging davon aus, dass die technisch-industrielle Modernisierung nach Modell der „Ersten" (westlich-kapitalistischen) oder der „Zweiten" (östlich-staatssozialistischen) Welt von den „Entwicklungsländern" nachzuholen sei, und dass durchaus auch ein Recht von Menschen darauf bestünde, dies zu tun. Zugleich war allerdings klar, dass die weltweite Ausdehnung *beider* Modelle noch mehr Umweltprobleme erzeugen würde (was sich bestätigt hat; vgl. Kap. 2). Die von den Vereinten Nationen eingesetzte „Weltkommission für Umwelt und Entwicklung" unter der

Leitung der Norwegerin Gro Harlem Brundtland (*1939) versuchte, diesen Antagonismus zu lösen und machte einen normativ sehr gehaltvollen Vorschlag. *Nachhaltige Entwicklung (Sustainable Development)* wurde der Ausdruck für umfassende globale (Umwelt)Gerechtigkeit:

> *„Nachhaltige[2] Entwicklung ist Entwicklung, die die Bedürfnisse der Gegenwart befriedigt, ohne zu riskieren, dass künftige Generationen ihre eigenen Bedürfnisse nicht befriedigen können. Zwei Schlüsselbegriffe sind wichtig: der Begriff von „Bedürfnissen", insbesondere der Grundbedürfnisse der Ärmsten der Welt, die die überwiegende Priorität haben sollten; und der Gedanke von Beschränkungen, die der Stand der Technologie und sozialen Organisation auf die Fähigkeit der Umwelt ausübt, gegenwärtige und zukünftige Bedürfnisse zu befriedigen." (UNCED 1987, S. 46)*

Entscheidend in normativer Hinsicht sind drei Aspekte, die sich bei der – leider – fast überall verbreiteten auf den ersten Satz verkürzten ‚Definition' nicht erschließen lassen.

- Erstens liegt ein *egalitäres Gerechtigkeitsprinzip* vor: Alle heutigen und künftigen Menschen sollen ihre Bedürfnisse befriedigen können. Bereits dies geht über die Idee kluger Ressourcennutzung weit hinaus.
- Zweitens wird ein weiteres Gerechtigkeitsprinzip hinzugefügt: die Ärmsten der Welt sollen *ausgleichende Gerechtigkeit* erfahren, weil deren Grundbedürfnisse prioritär gesetzt werden. Das ändert das erste Prinzip nicht, gibt aber für die Umsetzung eine klare Orientierung.
- Drittens geht es um die *begrenzte naturale und technische Basis* für menschliche Bedürfnisbefriedigung. Wo die Grenzen liegen, jenseits derer die Basis zerbröckelt, ist keine einfache Frage. Hier ist gerade die Humanökologie mit ihrer Expertise gefragt. Dennoch ist auch dieser Aspekt nicht ohne normative Positionierung zu behandeln. Wie groß soll die ‚Sicherheitsmarge' bei der Festlegung der Grenzen oder Leitplanken sein? Welche Folgen sollen wir bei einem Irrtum über die Grenzen akzeptieren, welche nicht? Auch hier ist die Ethik mit im Spiel, oft mittels des *Vorsorgeprinzips*, lieber auf der sicheren Seite zu irren, wenn irreversible und katastrophale Schäden zu erwarten sind (vgl. auch Jonas 1979). Solche Überlegungen richten sich auch gegen die Hybris im Denken, dass Menschen ‚alles' genau gestalten und kontrollieren könnten, was die Eigendynamiken (sozial-)ökologischer Systeme ignoriert.

---

[2]Im Original der deutschen Ausgabe steht „Dauerhafte" statt „Nachhaltige". Dies wurde später geändert, auch weil klar wurde, dass es nicht um ein temporal zu verstehendes Adjektiv geht, sondern dass Nachhaltige Entwicklung als *terminus technicus* einen komplexeren Bezugsrahmen herstellt.

Ethisch lassen sich die Prinzipien des „Brundtland"-Berichts aus der „Allgemeinen Erklärung der Menschenrechte" (Vereinte Nationen 1948) ableiten. Beides sind keine philosophischen Werke, dafür aber von den Vereinten Nationen beschlossene politische Konsensdokumente. Daher können die hier formulierten normativen Grundlagen als weitgehend von ‚Allen' akzeptiert und damit auch ethisch akzeptabel gelten. Zugleich wird allerdings deutlich, dass alles Nichtmenschliche tatsächlich lediglich als Ressource betrachtet wird, also *Naturdinge und Systeme nur einen Nutzwert für Menschen* haben. Zugleich wird betont, dass Menschen nur mit und in dafür geeigneten ökologischen Systemen leben können. Das ist mithin kein geringer Wert, sondern ganz im Gegenteil eine Notwendigkeit, die *conditio sine qua non*.

Einen weiteren Schritt in der politischen Ausbuchstabierung lieferte die UN-Konferenz für Umwelt und Entwicklung in Rio de Janeiro 1992. Sie macht in der umfassenden „Agenda 21" (Vereinte Nationen 1992) klar, dass die sozialen, die wirtschaftlichen, die Ressourcen- und die politischen Gestaltungsdimensionen zusammengedacht und -gestaltet werden müssen.[3] Die heute gültige „Agenda 2030" (United Nations 2015) ist eine aktualisierte Fortschreibung. Darüber hinaus wurden mit der Klimarahmenkonvention (UNFCCC 1992) und der Biodiversitätskonvention (CBD 1992) zwei entscheidende Handlungsfelder detailliert ausgeführt und seither in Folgekonferenzen und Vereinbarungen fortgeschrieben. Entscheidend ist nicht zuletzt, dass alle Signatarstaaten solcher UN-Vereinbarungen sich verpflichten, das Vereinbarte in nationales Recht umzusetzen. Das gilt auch für die Europäische Union. Das EU-Artenschutzrecht einschließlich Flora-Fauna-Habitat-Richtlinie und die verschiedenen Biodiversitätsstrategien wären sonst nicht entstanden. Auch hier vermittelt der politische Kompromiss eine zugrunde liegende geteilte ethische Wertschätzung.

Einen neuen Wertaspekt hat die Biodiversitätskonvention ins Spiel gebracht: Sie formuliert im ersten Satz der Präambel einen eigenen Wert (*intrinsic value*, CBD 1992, preamble) der biologischen Vielfalt, die sich auf die Ebenen der genetischen, der Arten- und der Ökosystemvielfalt bezieht.[4] Die, zudem oft ökonomisch verkürzte, Konzentration auf den alleinigen Nutzwert für Menschen mit der moralischen Nichtberücksichtigung nichtmenschlichen Lebewesen hatte ja mit

---

[3]Die später – leider – oft erfolgte Aufteilung in getrennte ‚Säulen' ist ebenso falsch wie nicht zielführend, weil es eben genau um integrierte sozial-ökologische Systeme geht.

[4]Das Bundesnaturschutzgesetz hat dies seit der Fassung von 2002 übernommen, ebenso die meisten Landesnaturschutzgesetze. Was genau dieser „eigene Wert" bedeutet, bleibt in der CBD und den Gesetzen unspezifiziert. Klar ist jedoch, dass er über klugheitsethische Aspekte des aufgeklärten Eigeninteresses von Menschen ausdrücklich hinausgeht.

zur gegenwärtigen Biodiversitätskrise geführt. Der *intrinsic value* adressiert nicht zuletzt dieses Problem.

4.4 Werte (in) der Natur und ihre Begründung – eine Übersicht

Schon lange ist die ethische Debatte in voller Fahrt, welche Werte die Natur, die Biosphäre, Ökosysteme, die Biodiversität oder bestimmte Gruppen von Lebewesen haben – eng verbunden mit der politischen Frage, wie entsprechend gehandelt werden sollte. Systematisch gesehen lassen sich drei Wertdimensionen unterscheiden (Eser und Potthast 1999):

1. Der *instrumentelle Wert* ist ein *Nutzen- und Funktionswert für etwas oder jemand Anderes;* er ist als Nutzen/Funktion tendenziell objektiv erfassbar. Wie oben ausgeführt, sind natürliche Ressourcen und deren Systeme sogar essentiell für das gute Leben von Menschen. Während ein einfaches Werkzeug, wenn es kaputt geht, leicht durch ein anderes ersetzt werden kann, gilt dies eben nicht für das *Klimasystem,* die *Biosphäre* und einzelne *Spezies.* Biologische Vielfalt und die meisten Ökosystem(dienst)leistungen sind technisch *nicht substituierbar,* daher müssen sie im Sinne Nachhaltiger Entwicklung erhalten werden (Konzeption „starker Nachhaltigkeit", Ott und Döring 2011); dies impliziert eine umfassende ethische Verpflichtung zu ihrer Erhaltung.
2. Der *relationale Wert* oder *Eigenwert* oder *eudaimonistische Wert* drückt eine nicht (rein) instrumentelle Beziehung zwischen Menschen und nichtmenschlichen Dingen oder Wesen aus und befördert so das gute (moralisch anzustrebende, eudaimonistische) Leben. Dies kann bei unterschiedlichen Menschen unterschiedlich sein, weil es um persönliche Erlebnisse geht, um Ästhetik und kulturell unterschiedliche Wertschätzungen. Solche Werte sind intersubjektiv nachvollziehbar; sie verpflichten indirekt, aus Respekt Anderen gegenüber. Die nicht zuletzt kommunikative Bedeutung relationaler Werte wird derzeit stark betont (Pascual et al. 2018).
3. Der *Selbstwert* oder *intrinsische Wert* oder *Wert an und für sich* entspringt dagegen nicht subjektiver Wertschätzung, sondern liegt in den Dingen oder Wesen selbst. Er entspringt einer kriterienbasierten ethischen Setzung, die unterschiedlich ausfallen kann: In der *Anthropozentrik* sind dies *alle Menschen* aufgrund ihrer Selbstzwecklichkeit als vernunftbegabte Wesen *und* als Teil der Menschheit, unabhängig von individuellen aktuellen Fähigkeiten. In der *Pathozentrik* bzw. dem *Sentientismus* sind es alle Wesen, die Schmerz- bzw.

Empfindungen (selbst)bewusst wahrnehmen können; in der *Biozentrik* sind alle Lebewesen qua ihres Lebensdrangs *direkt* moralisch bedeutsam (Teil der *moral community,* s. u.), und im *Holismus* besitzen auch Unbelebtes und Einheiten wie Arten oder Ökosysteme einen Selbstzweck und daher Selbstwert. Selbstwerte als absolute Werte sind grundsätzlich nicht gradier- und abwägbar. Dennoch wird diskutiert, ob sich eine moralische Bedeutsamkeit nichtmenschlicher Wesen (und Dinge) doch ggf. sekundär abstufen ließe (gradualistische Positionen; taxon- bzw. speziesspezifischer Selbstwert).

Begründungs- und Erkenntnisfragen zum letzten Punkt werden intensiv diskutiert, es sind dabei unterschiedliche Ebenen zu beachten:

- Diskurs- und Erkenntnisebene: Es sind stets Menschen als moralfähige und moralpflichtige Wesen *(moral agents),* die über solche Fragen kommunizieren – *epistemischer Anthropozentrismus bzw. epistemische Anthroporelationalität* ist unhintergehbar.
- Ob moralische Werte und ggf. Normen nur aus der Sphäre des Menschen stammen können, oder sie auch andere Herkünfte haben können, ist eine davon verschiedene Frage. Die *Herkunft von Werten* wird zuweilen auch nicht anthropogenen *Quellen* (Götter/Geister, die nichtmenschliche Natur) zugeschrieben.
- Die *moralische Anthropozentrik* ist die Position, die nur Menschen einen Selbstwert zuschreibt und daher als direkt zu berücksichtigende *moral community* sieht. Andere Positionen werden – trotz erheblicher Unterschiede im Umfang der *moral community* (s. o.) – als *Physiozentrik* zusammengefasst. Hierzu gehören auch zunehmend international beachtete nichtwestliche Ansätze wie *Ubuntu* aus Subsahara-Afrika und *Patcha Mama* (Mutter Erde) aus Südamerika, die die spirituelle Gemeinschaft aller Wesen (in) der Natur betonen.

4.5   Integrative ethisch-humanökologische Perspektiven

Für die Humanökologie als Wissenschaft gilt die methodologische Setzung, dass sie in ihren *Erkenntnissen* frei von persönlichen Wertsetzungen und weltanschaulichen Annahmen bleiben muss.[5] Als geteilte normative Basis ‚reicht' es,

---

[5]Dies gilt aber nicht für die Motivation der Forschenden oder die Wahl der Themen!

auf Basis des vorhandenen humanökologischen Wissens und unter Anerkennt-
nis der Prinzipien Nachhaltiger Entwicklung sich zu Fragen der Klima- oder der
Biodiversitätskrise zu positionieren, ohne die Debatten um Selbstwerte (in) der
Natur entscheiden zu müssen. Die mit der Humanökologie verbundene Ethik
muss notwendig interdisziplinär sein, weil sie ethische Urteile nur im Hori-
zont bestimmter empirischer Informiertheit angemessen treffen kann; sie muss
darüber hinaus transdisziplinär sein, um auch mit außerakademischen Akteurs-
gruppen das moralische Gespräch zu suchen. Dabei verfällt sie genau nicht
Sein-Sollens- oder Naturalistischen Fehlschlüssen, sondern kommt zu nachvoll-
ziehbaren begründeten „gemischten Urteilen" (Potthast 2015). In Anschluss an
die Debatten um Nachhaltige Entwicklung ist dabei klar: Sowohl die Strategie
der gesteigerten *Effizienz* menschlicher Ressourcennutzung als auch die Strategie
der *Resilienz* bzw. *Konsistenz* mit der Verwendung nachwachsender, recycling-
fähiger und ‚naturverträglicher' Materialien und Prozesse sind zwar notwenige,
aber nicht hinreichende Ansätze. Unumgänglich ist zudem vor allem die Strategie der *Suffizienz,* der Reduktion des absoluten Ressourcenverbrauchs, wie sie der
Agro-Urban-Industrielle Komplex derzeit mit sich bringt. Dies erfordert grundlegende gesellschaftliche, sozioökonomische und kulturelle Transformationen aller
Länder, die diesem Komplex (bislang und ggf. schon lange) zugehören. Im Sinne
der Generationengerechtigkeit lassen sich von solchen Ländern auch verstärkte
Anstrengungen im Biodiversitäts- und Klimaschutz begründet fordern (Ott und
Döring 2011; United Nations 2015).

Die in Kap. 4.4 geschilderten Aspekte setzen eine oft binär verstandene
Trennung zwischen Menschen und Natur voraus, die aber, wie erwähnt, frag-
lich ist (Herrmann 2016). Mithin ist auch mehr „*relationale"* oder „*inklusive
Ethik"* (Eser und Potthast 1999) angezeigt, die jenseits des ‚entweder-oder' ope-
riert. Werte und Verpflichtungen, die durch die Interaktion von Menschen in
und mit (sozial-)ökologischen Systemen bestehen, sollten im Fokus liegen. Dann
erweisen sich scheinbare Konflikte zwischen Werten „des Menschen" und Selbst-
werten „der Natur" zumeist als Konflikte unterschiedlicher Formen menschlicher
Naturinteraktionen. Dabei ist die moralische Position einer auch nutzenunabhän-
gig bestehenden Wertschätzung der und Verantwortung für die nichtmenschliche
Mitwelt in (sozial-)ökologischen Systemen durchaus plausibel.

Die Entstehung einer differenzierten menschlichen Kultur ist als Folge von Aneignungs- und Ausbeutungsstrategien natürlicher Ressourcen zu sehen, aus der eine Weltaneignung entspringt, die von den Ideen zu deren Umsetzung fortschreitet und in diesem Prozess ihre eigenen Bedeutungsysteme hervorbringt.

Eine Kultur *ist damit* **das** *menschliche Ökosystem,* das gleichsam als ökosystemische Suprastruktur über den von Menschen genutzten oder beeinflussten Ökosystemen liegt. Humanökologie befasst sich entsprechend mit dem Gesamt menschlicher Ideen und Handlungen, soweit sie ökologisch für Menschen wie für übrige Organismen wirksam sind oder sein können. Sie befasst sich nicht etwa ‚mit Allem', sondern konzentriert sich auf die Voraussetzungen sowie die Folgen und Nebenfolgen für den ökologischen Zustand des Erdsystems und seiner Teile, soweit sie in der Verantwortung von Menschen und ihren Wirkungsmöglichkeiten liegen. Sie ist somit dem von den Vereinten Nationen beschlossenen normativen Prinzip einer Nachhaltigen Entwicklung verpflichtet.

Humanökologie thematisiert Wechselbeziehungen zwischen einerseits den biologischen Grundbedürfnissen und kulturell moderierten Lebensansprüchen der Art *Homo sapiens* und andererseits den von Menschen genutzten Ökosystemen bzw. der Biosphäre. In der Formulierung des MEA (2005) stehen dabei die ‚Dienstleistungs'-Ansprüche von Menschen im Vordergrund. Deren Folgen und Nebenfolgen für die von Menschen genutzten oder beeinflussten Habitate und Ökosysteme sind weitreichend. Sie sind seit dem Beginn der Kolonialisierung, der Abschöpfung externer Ökosysteme, zu einer Bedrohung der Lebensgrundlagen für Menschen selbst und vieler anderer Organismen angewachsen.

Die Humanökologie fokussiert auf Themen der Bevölkerungsforschung und auf die Problematik der Mensch-genutzten Ressourcen und ihrer Verfügbarkeit. Zwangsläufig ergeben sich daraus Analysen über die Belastungen von

Ökosystemen durch menschliche Nutzung und Schlussfolgerungen über die Notwendigkeiten und Möglichkeiten zur Sicherung und Wiederherstellung ökologischer Systeme. Offenkundig ist die Erörterung dieser Gegenstände nicht nur an natur- und lebenswissenschaftlich gefundene Sachverhalte gebunden, sondern auch an gesellschaftlich konstruierte Wirklichkeitsvorstellungen und ihren zugrunde liegenden Wertvorstellungen.

Humanökologie wird zwar notwendig aus epistemisch-anthropozentrischer Perspektive betrieben, verfolgt aber aufgrund der gewonnen Einsicht, dass Menschen der Natur unterschiedliche Wertdimensionen beimessen (auch eudaimonistisch-relationale und z.T. Selbstwerte), nicht nur Ziele des bloßen Nutzenwertes. Nichtmenschlichen Organismen und Systemen wird auch ein eigener Wert zugesprochen, der Menschen zu ihrem Schutz verpflichtet. Insofern ist es Ziel humanökologisch-ethischer Reflexionen, Begründungen und konkrete Wege für eine Nachhaltige Entwicklung, für ein gelingendes Zusammenleben von Menschen in und mit anderen Lebensformen und Umweltsystemen zu erkunden und zu begründen.

Entsprechend subsumiert der Begriff „Humanökologie" drei Ökologiekonzepte, die sich unterschiedlicher Akzentuierungen des Ökologiebegriffs verdanken:

- einem naturwissenschaftlich-lebenswissenschaftlichen Ökologieverständnis,
- einem sozialwissenschaftlich moderiertem Ökologieverständnis und
- einem ökologisch sowohl orientierten als auch orientierenden Ethikverständnis.

Entsprechend unterschiedlich und nuancenreich fallen im Schrifttum Definitionen von „Humanökologie" aus. Zusammenfassend erschließt sich die Humanökologie als Theorie des Wechselspiels zwischen Gesellschaft und Umwelt, Mensch und Natur in folgenden Bezugslinien und Horizonten:

- durch problemorientierte Forschungsfragen, die politikrelevant sind und die Betroffenen aus der lebensweltlichen Praxis einbeziehen,
- mit inter- und transdisziplinärem Bezug, der die Methodenfrage neu stellt, auch unter Einbeziehung lokaler Kompetenzen *(indigenous knowledge),*
- im partnerschaftlichen Austausch, Globaler Süden und Globaler Norden eingeschlossen.

Was Sie aus diesem *essential* mitnehmen können

- Die „neolithischen Revolutionen" markieren mit der Sesshaftwerdung von Menschen und der perspektivischen Erfindung der Tierwirtschaft den Anfang des *Agro-Urban-Industriellen* Komplexes und damit des heute prekären Zustands des globalen Ökosystems.
- Die Humanökologie bietet als synthetische Wissenschaft die Möglichkeit zur ganzheitlichen Analyse dieses Zustands.
- Sie entwickelt partizipatorische und umweltverträgliche Lösungskonzepte.
- Sie begründet einen ethisch verantwortlichen nachhaltigen und naturerhaltenden Umgang mit der Umwelt

B. Herrmann et al., *Humanökologie*, essentials, https://doi.org/10.1007/978-3-658-32983-9

Literatur

Kapitel 1

Cassirer, E. (1996). *Versuch über den Menschen. Einführung in eine Philosophie der Kultur.* Hamburg: Meiner.

Cassirer, E. (2001). *Gesammelte Werke.* Hamburger Ausgabe, hrsg. von Recki B. Darmstadt: Wissenschaftliche Buchgesellschaft.

Descola, P. (2013). *Jenseits von Natur und Kultur. Taschenbuch Wissenschaft* (Bd. 2076). Frankfurt a. M.: Suhrkamp.

Elkana, Y. (2006). A theatre for the enactment of the anthropology of knowledge. In D. Grimm & R. Meyer-Kalkus (Hrsg.), *25 Jahre Wissenschaftskolleg zu Berlin 1981–2006* (S. 127–139). Berlin: Akademie.

FAO. (2007). *Food and agriculture organization of the United Nations. The state of the world's animal genetic resources for food and agriculture. Commission on genetic resources for food and agriculture.* Rom: FAO.

FAO. (2010). *Food and agriculture organization of the United Nations. Second report on the state of the world's plant genetic resources for food and agriculture. Commission on genetic resources for food and agriculture.* Rom: FAO.

FAO. (2020). *State of the world series.* Rom: FAO. http://www.fao.org/publications/en/.

Frühwald, W. (2006). Wirkungen der Freiheit. In D. Grimm & R. Meyer-Kalkus (Hrsg.), *25 Jahre Wissenschaftskolleg zu Berlin 1981–2006* (S. 1–27). Berlin: Akademie.

Hartmann, N. (1980). *Philosophie der Natur. Abriß der speziellen Kategorienlehre* (2. Aufl.). Berlin: De Gruyter.

Herrmann, B. (2016). *Umweltgeschichte. Eine Einführung in Grundbegriffe* (2. Aufl.). Berlin: Springer Spektrum.

Herrmann, B. (2021). Die Entdeckung der Umwelt. *Saeculum 71*(1), im Druck.

Herrmann, B., & Sieglerschmidt, J. (2018). *Umweltgeschichte und Kausalität. Entwurf einer Methodenlehre.* Springer Spectrum essentials. https://doi.org/10.1007/978-3-658-20921-6.

IPBES Intergovernmental Platform on Biodiversity and Ecosystem Services. (2019). Summary for policymakers of the global assessment report on biodiversity and ecosystem services of the Intergovernmental Science-Policy Platform on Biodiversity and Ecosystem Services. https://ipbes.net/sites/default/files/ipbes_7_10_add.1_en_1.pdf.

IPCC. (2014). Klimaänderung 2014: Synthesebericht. Beitrag der Arbeitsgruppen I, II und III zum Fünften Sachstandsbericht des Zwischenstaatlichen Ausschusses für Klimaänderungen (IPCC). In R. Pachauri & L. Meyer (Hrsg.), IPCC, Genf, Schweiz. Deutsche Übersetzung durch Deutsche IPCC-Koordinierungsstelle, Bonn, 2016. https://www.de-ipcc.de/media/content/IPCC-AR5_SYR_barrierefrei.pdf.

James, W. (2012). *Pragmatismus. Ein neuer Name für einige alte Denkweisen* (Übers .u. hrsg. von Schubert K, Spree A.). Hamburg: Meiner.

MEA Millennium Ecosystem Assessment. (2005). *Ecosystems and human well-being: synthesis*. Washington: Island Press. http://www.millenniumassessment.org/documents/document.356.aspx.pdf.

Uexküll, J. v. (2014). *Umwelt und Innenwelt der Tiere* (Hrsg. von Mildenberger F, Herrmann B.). Berlin: Springer Spektrum.

Kapitel 2

Berger, R., & Ehrendorfer, F. (Hrsg.). (2011). *Ökosystem Wien. Die Naturgeschichte einer Stadt*. Wien: Böhlau.

Brunner, K., & Schneider, P. (Hrsg.). (2005). *Umwelt Stadt. Geschichte des Natur und Lebensraumes Wien*. Wien: Böhlau.

Cavalli-Sforza, L., Menozzi, P., & Piazza, A. (1994). *The history and geography of human genes*. Princeton: Princeton University Press.

de Vries, J., & van der Woude, A. (1997). *The first modern economy: Success, failure, and perseverance of the Dutch economy 1500–1815*. Cambridge: Cambridge University Press.

Drenckhahn, D., Arneth, A., Filser, J., Haberl, H., Hansjürgens, B., Herrmann, B., Leuschner, C., Mosbrugger, V., Reusch, T., Schäffer, A., Scherer-Lorenzen, M., & Tockner, K. (2020) Online-Dokumentationsband zu Leopoldina, Diskussion Nr. 24: Globale Biodiversität in der Krise – Was können Deutschland und die EU dagegen tun? https://www.leopoldina.org/uploads/tx_leopublication/2020_Dokumentationsband_Biodiversitaetskrise.pdf.

Durham, W. (1991). *Genes, culture, and human diversity*. Stanford: Stanford University Press.

FAO. (2010). *Food and agriculture organization of the United Nations. Second report on the state of the world's plant genetic resources for food and agriculture. Commission on genetic resources for food and agriculture*. Rom: FAO.

FAO. (2007). *Food and Agriculture Organization of the United Nations. The state of the world's animal genetic resources for food and agriculture. Commission on genetic resources for food and agriculture*. Rom: FAO.

Groh, D. (1992). Strategien, Zeit und Ressourcen. Risikominimierung, Unterproduktivität und Mußepräferenz – Die zentralen Kategorien von Subsistenzökonomien. In: D. Groh (Hrsg.). *Anthropologische Dimensionen der Geschichte* (S. 54–113). Frankfurt: Suhrkamp.

Henke, W., & Rothe, H. (2014). *Menschwerdung*. Frankfurt a. M.: S. Fischer.

Herrmann, B. (2016). *Umweltgeschichte. Eine Einführung in Grundbegriffe* (2. Aufl.). Berlin: Springer Spektrum.

Herrmann, B. (2019). *Das menschliche Ökosystem*. Wiesbaden: Springer Spektrum (essential). https://doi.org/10.1007/978-3-658-24943-4.

Isenberg, A. (Hrsg.). (2017). *The Oxford handbook of environmental history*. Oxford: Oxford University Press.

IPBES Intergovernmental Platform on Biodiversity and Ecosystem Services. (2019). Summary for policymakers of the global assessment report on biodiversity and ecosystem services of the Intergovernmental Science-Policy Platform on Biodiversity and Ecosystem Services. https://ipbes.net/sites/default/files/ipbes_7_10_add.1_en_1.pdf.

IPCC. (2014). Klimaänderung 2014: Synthesebericht. Beitrag der Arbeitsgruppen I, II und III zum Fünften Sachstandsbericht des Zwischenstaatlichen Ausschusses für Klimaänderungen (IPCC) [Pachauri R, Meyer L (Hrsg.)]. IPCC, Genf, Schweiz. Deutsche Übersetzung durch Deutsche IPCC-Koordinierungsstelle, Bonn, 2016. https://www.de-ipcc.de/media/content/IPCC-AR5_SYR_barrierefrei.pdf.

Kirchhoff, T. (2015). Die Zeitform der Entwicklung von Ökosystemen und ökologischen Gesellschaften. Richtschnur für menschliche Vergesellschaftung? In: G. Hartung (Hrsg.), *Mensch und Zeit* (S. 226–248). Wiesbaden: Springer.

McNeill, J. (2005). *Blue Planet. Die Geschichte der Umwelt im 20. Jahrhundert*. Schriftenreihe Bundeszentrale für Politische Bildung Bd 518, Bonn.

MEA Millennium Ecosystem Assessment. (2005). *Ecosystems and human well-being: Synthesis*. Washington: Island Press. http://www.millenniumassessment.org/documents/document.356.aspx.pdf.

Moran, E. (2008). *Human adaptability. An introduction to ecological anthropology* (3. Aufl.). Boulder: Westview Press.

Reich, D. (2018). *Who we are and how we got here*. Oxford: Oxford University Press.

Robinson, J., & Wiegand, K. (Hrsg.). (2008). *Die Ursprünge der modernen Welt. Geschichte im wissenschaftlichen Vergleich*. Frankfurt a. M.: Fischer Taschenbuch.

Sahlins, M. (1974). *Stone age economics*. London: Tavistock Publ.

Schaefer, M. (2012). *Wörterbuch der Ökologie* (5. Aufl.). Heidelberg: Spektrum Akademischer.

Schutkowski H (2006) *Human Ecology. Biocultural adaptations in human communities* (Ecological Studies 182). Heidelberg: Springer.

Sieferle, R., Krausmann, F., Schandl, H., & Winiwarter, V. (2006). *Das Ende der Fläche. Zum gesellschaftlichen Stoffwechsel der Industrialisierung*. Böhlau: Köln.

Sukopp, H., & Wittig, R. (Hrsg.). (1998). *Stadtökologie* (2. Aufl.). Stuttgart: Gustav Fischer.

TEEB. (2008). The economics of ecosystems and biodiversity. European Communities. http://www.teebweb.org/wp-content/uploads/Study%20and%20Reports/Additional%20Reports/Interim%20report/TEEB%20Interim%20Report_English.pdf.

WBGU. (2000). *Wissenschaftlicher Beirat der Bundesregierung Globale Umweltveränderungen, Jahresgutachten 1999: Welt im Wandel. Erhaltung und nachhaltige Nutzung der Biosphäre*. Berlin: Springer. https://www.wbgu.de/fileadmin/user_upload/wbgu/publikationen/hauptgutachten/hg1999/pdf/wbgu_jg1999.pdf.

Weber, M. (1988). Die „Objektivität" sozialwissenschaftlicher und sozialpolitischer Erkenntnis. In J. Wickelmann (Hrsg.), *Weber M, Gesammelte Aufsätze zur Wissenschaftslehre* (7. Aufl., S. 146–214). Tübingen: Mohr.

Weitere Lesempfehlungen

Cox, C. B., Moore, P., & Ladle, R. (2020). *Biogeography: An ecological and evolutionary approach* (10. Aufl.). Hoboken: Wiley.

Freye, H.-A. (1986). *Einführung in die Humanökologie für Mediziner und Biologen.* Heidelberg: Quelle & Meyer.

Goudie, A. (2013). *The human impact on the natural environment* (7. Aufl.). Chichester: Wiley Blackwell.

Nentwig, W. (2005). *Humanökologie* (2. Aufl.). Berlin: Springer.

Thomas, W. (Hrsg.). (1956). *Man's role in changing the face of the earth.* Chicago: University of Chicago Press.

Turner, B., Clark, W., Kates, R., Richard, J., Mathews, J., & Meyer, W. (Hrsg.). (1990). *The earth as transformed by human action.* Cambridge: Cambridge University Press.

Kapitel 3

Borden, R., & Jacobs, J. (Hrsg.). (1989). *International directory of human ecologists* (2. Aufl.). Bar Harbor (USA): Society for Human Ecology & College of the Atlantic.

Brislin, R. (1980). Cross-cultural research methods. In I. Altmann, A. Rapoport, J. Wohlwill (Hrsg.), *Human behavior and environment. Vol. 4: Environment and culture* (S. 47–82). New York: Plenum Press.

Bruckmeier, K. (2004). Die unbekannte Geschichte der Humanökologie. In W. Serbser (Hrsg.), *Humanökologie* (S. 45–120). München: Oekom.

Bruckmeier, K. (2005). Das schwedische Forschungsprogramm „SUCOZOMA–Sustainable Coastal Zone Management". Erfahrungen und Ergebnisse. In B. Glaeser (Hrsg.), *Küste, Ökologie und Mensch* (S. 55–97). München: Oekom.

Buttel, F. (1987). New directions in environmental sociology. *American Review of Sociology, 13,* 456–488.

Cafaro, P. (2001). Thoreau, Leopold, and Carson: Toward an environmental virtue ethics. *Environmental Ethics, 23*(1), 3–17.

Catton, W. (1993). Sociology as an ecological science. In S. Wright, et al. (Hrsg.), *Human Ecology: Crossing boundaries* (S. 74–86). Fort Collins (USA): Society for Human Ecology.

Corbin, A. (1994). *Meereslust. Das Abendland und die Entdeckung der Küste.* Frankfurt a. M.: Fischer.

Dunlap, R., & Catton, W. (1979). Environmental sociology. *Annual Review of Sociology, 79*(5), 243–273.

Egger, K., & Glaeser, B. (1975). *Politische Ökologie der Usambara-Berge in Tanzania.* Bensheim: Kübel-Stiftung.

Fischer, L., & Reise, K. (Hrsg.). (2011). *Küstenmentalität und Klimawandel. Küstenwandel als kulturelle und soziale Herausforderung.* München: Oekom.

Glaeser, B. (1992). Natur in der Krise? Ein kulturelles Missverständnis. *GAIA, 1*(4), 195–203.

Glaeser, B. (1996). Sociology of the environment: A German-American comparison. *Human Ecology Review, 3*(1), 32–42.

Glaeser, B. (1996b). Humanökologie: Der sozialwissenschaftliche Ansatz. *Naturwissenschaften, 83*(4), 145–152.

Glaeser, B. (1999). Integrated Coastal Zone Management in Sweden: Assessing conflicts to attain sustainability. In W. Salomons, R. Turner, L. Lacerda, & S. Ramachandran (Hrsg.), *Integrated coastal zone management* (S. 355–375). Berlin: Springer.

Glaeser, B. (2004). Humanökologie im internationalen Kontext. In W. Serbser (Hrsg.), *Humanökologie* (S. 25–44). München: Oekom.

Glaeser, B. (Hrsg.). (2005). *Küste, Ökologie und Mensch. Integriertes Küstenmanagement als Instrument nachhaltiger Entwicklung.* München: Oekom.

Glaeser, B. (2006). Nachhaltigkeit in Forscherverbünden: Ein Thema für Humanökologen. In B. Glaeser (Hrsg.), *Fachübergreifende Nachhaltigkeitsforschung* (S. 17–37). München: Oekom.

Glaeser, B., Gee, K., Kannen, A., & Sterr, H. (2004). Auf dem Weg zur nationalen Strategie im integrierten Küstenzonenmanagement: Raumordnerische Perspektiven. Informationen zur Raumentwicklung 7–8. *BBR, 2004,* 505–513.

Glaeser, B., Kannen, A., & Schernewski, G. (2007). Unterstützung für ein Integriertes Küstenzonenmanagement: Netzwerke und Forschungsverbünde in Nord- und Ostsee. Informationen zur Raumentwicklung 5. *BBR, 2007,* 297–305.

Glaser, M., Krause, G., Ratter, B., & Welp, M. (Hrsg.). (2012). *Human-nature interactions in the Anthropocene* (S. 193–222). New York: Routledge.

Glaser, M., Radjawali, I., Ferse, S., & Glaeser, B. (2010). Nested participation in hierarchical societies? Lessons for social-ecological research and management. *International Journal of Society Systems Science, 2*(4), 390–414.

Haeckel, E. (1866). *Generelle Morphologie der Organismen* (Bd. 2). Berlin: Reimer.

Hawley, A. (1986). *Human ecology. A theoretical essay.* Chicago: University of Chicago Press.

Kannen, A., Schernewski, G., Krämer, I., Lange, M., Janßen, H., Stybel, N. (Hrsg.) (2010). Forschung für ein Integriertes Küstenzonenmanagement. Coastline Reports 2010/15.

Knötig, H. (Hrsg.). (1976). *Internationale Tagung für Humanökologie 1975* (Bd. 2). St. Saphorin (Schweiz): Humanökologische Gesellschaft.

König, R. (Hrsg.) (1958). *Soziologie. Das Fischer Lexikon.* Frankfurt a. M.: Fischer.

McKenzie, R. (1968). *On human ecology* (S. 33–48). Chicago: University of Chicago Press.

Meadows, D., Meadows, D., Zahn, E., & Milling, P. (1972). *Die Grenzen des Wachstums. Bericht des Club of Rome zur Lage der Menschheit.* Stuttgart: DVA.

Merchant, C. (1980) *The death of nature. Women, ecology, and the scientific revolution.* New York: Harper and Row.

Meyer-Abich, K. (1972). Die ökologische Grenze des Wirtschaftswachstums. *Umschau, 72*(20), 645–649.

Moebius, K. (1877) *Die Auster und die Austernwirthschaft.* Berlin: Wiegandt, Hempel & Parey.

Morrow, R., & Brown, D. (1994). *Critical theory and methodology.* Thousand Oaks: Sage.

National Research Council. (1995). *Science, policy, and the coast. Improving decisionmaking.* Washington D.C.: National Academy Press.

Paine, L. (2013). *The Sea and civilization. A maritime history of the world.* New York: Knopf.

Park, R. (1952). *Human communities: The city and human ecology.* Glencoe (USA): Free Press.

Park, R., Burgess, E., & McKenzie, R. (1925). *The city*. Chicago: University of Chicago Press.

Rappaport, R. (1968). *Pigs for the ancestors*. New Haven: Yale University Press.

Reise, K. (1980). Hundert Jahre Biozönose. Die Evolution eines ökologischen Begriffes. *Naturwissenschaftliche Rundschau, 33,* 328–335.

Reise, K. (1998). Einstige Austernbänke. In: Landesamt für den Nationalpark Schleswig-Holsteinisches Wattenmeer und Umweltbundesamt (Hrsg.) *Umweltatlas Wattenmeer. Band 1: Nordfriesisches und Dithmarscher Wattenmeer*. Stuttgart: Ulmer.

Rich, N. (5. August 2018). Losing earth. *New York Times Magazine*.

Schnaiberg, A. (1980). *The environment. From surplus to scarcity*. New York: Oxford University Press.

Scholtz, G. (2016). *Philosophie des Meeres*. Hamburg: Mare.

Serbser, W. (Hrsg.). (2004). *Humanökologie: Ursprünge, Trends, Zukünfte*. München: Oekom.

Steward, J. (1955). *Theory of culture change. The methodology of multilinear evolution*. Urbana: University of Illinois Press.

Suzuki, S., Borden, R., & Hens, L. (Hrsg.). (1991). *Human ecology-coming of age: An international overview*. Brussels: VUB-Press.

Teherani-Krönner, P., & Glaeser, B. (2020). Human-. *Kultur- und Ethnoökologie. Natur und Landschaft, 95*(9–10), 407–417.

Kapitel 4

Bruckmeier, K., & Serbser, W. H. (Hrsg.). (2013). *Ethik und Umweltpolitik. Humanökologische Positionen und Perspektiven*. München: Oekom.

Butzer, K. W. (Hrsg) (1992). The Americas before and after 1492. *Annals of the Association of American Geographers, 82*(3), 343–565.

Carlowitz, H. C. v. (1713). *Silvicultura oeconomica – Oder hausswirthschaftliche Nachricht und naturmässige Anweisung zur wilden Baum-Zucht. Joh. Fried*. Leipzig: Braun, Digital. http://reader.digitale-sammlungen.de/resolve/display/bsb10214444.html.

CBD. (1992). *Convention on biological diversity*. New York: United Nations. https://www.cbd.int/convention/text/.

Cassirer, E. (1996). *Versuch über den Menschen. Einführung in eine Philosophie der Kultur*. Hamburg: Meiner.

Cronon, W. (Hrsg.). (1995). *Uncommon ground: Rethinking the human place in nature*. New York: Norton.

Descola, P. (2013). *Jenseits von Natur und Kultur*. Frankfurt a. M.: Suhrkamp.

dt.: Hauff, V. (Hrsg.) (1987). *Unsere gemeinsame Zukunft. Der Brundtland-Bericht der Weltkommission für Umwelt und Entwicklung*. Greven: Eggenkamp.

Eser, U., & Potthast, T. (1999). *Naturschutzethik. Eine Einführung für die Praxis*. Baden-Baden: Nomos.

Herrmann, B. (2016). *Umweltgeschichte. Eine Einführung in Grundbegriffe* (2. Aufl.). Heidelberg: Springer.

Jonas, H. (1979). *Das Prinzip Verantwortung. Versuch einer Ethik für die technologische Zivilisation*. Frankfurt a. M.: Suhrkamp.

Meadows, D., Meadows, D., Zahn, E., & Milling, P. (1972). *Die Grenzen des Wachstums. Bericht des Club of Rome zur Lage der Menschheit*. Stuttgart: DVA.

Merchant, C. (2020). *Der Tod der Natur. Ökologie, Frauen und neuzeitliche Naturwissenschaft* (3. Aufl.). München: Oekom (engl. Orig. 1980).

Ostrom, E. (1999). *Die Verfassung der Allmende. Jenseits von Staat und Markt.* Tübingen: Mohr Siebeck (engl. Orig. 1980).

Ott, K., Dierks, J., & Voget-Kleschin, L. (Hrsg.). (2017). *Handbuch Umweltethik.* Stuttgart: Metzler.

Ott, K., & Döring, R. (2011). *Theorie und Praxis starker Nachhaltigkeit* (3. Aufl.). Marburg: Metropolis.

Pascual, U., Gould, R., & Chan, K. M. A. (Hrsg.) (2018). Sustainability challenges: Relational values. *Current Opinion in Environmental Sustainability, 35,* 1–132. https://www.scienc edirect.com/journal/current-opinion-in-environmental-sustainability/vol/35.

Potthast, T. (2006). Naturschutz und Naturwissenschaft – Symbiose oder Antagonismus? Zur Beharrung und zum Wandel prägender Wissensformen vom ausgehenden 19. Jahrhundert bis in die Gegenwart. In H. W. Frohn & F. Schmoll (Hrsg.), *Natur und Staat. Staatlicher Naturschutz in Deutschland 1906–2006* (S. 343–444). Bonn: Bundesamt für Naturschutz.

Potthast, T. (2015). Ethics and sustainability science beyond Hume, Moore and Weber – Taking epistemic-moral hybrids seriously. In S. Meisch, J. Lundershausen, L. Bossert, & M. Rockoff (Hrsg.), *Ethics of science in the research for sustainable development* (S. 129–152). Baden-Baden: Nomos.

Schmoll, F. (2004). *Erinnerung an die Natur. Die Geschichte des Naturschutzes im deutschen Kaiserreich.* Frankfurt a. M.: Campus.

Shtilmark, F. (2003). *History of the russian Zapovedniks, 1895–1995.* Edinburgh: Russian Nature Press (russ. Orig. 1996).

UNCED – United Nations Conference on Environment and Development. (1987). *Our common future.* New York: United Nations. http://www.un-documents.net/our-common-fut ure.pdf.

United Nations. (2015). *The sustainable development goals.* New York: United Nations. https://sustainabledevelopment.un.org/sdgs.

UNFCCC. (1992). *Rahmenabkommen der Vereinten Nationen über Klimaveränderungen.* New York: United Nations. https://unfccc.int/resource/docs/convkp/convger.pdf.

Vereinte Nationen. (1948). Die Allgemeine Erklärung der Menschenrechte. United Nations, New York https://www.ohchr.org/EN/UDHR/Pages/Language.aspx?LangID=ger.

Vereinte Nationen. (1992). Agenda 21. Konferenz der Vereinten Nationen für Umwelt und Entwicklung Rio de Janeiro. United Nations, New York https://www.un.org/Depts/ger man/conf/agenda21/agenda_21.pdf.

Kapitel 5

MEA Millennium Ecosystem Assessment. (2005). *Ecosystems and human well-being: Synthesis.* Washington: Island Press. http://www.millenniumassessment.org/documents/doc ument.356.aspx.pdf.

Printed in the United States
by Baker & Taylor Publisher Services

A STUDY ON AU

A STUDY ON AUTHORITY

Herbert Marcuse

Translated by Joris De Bres

VERSO

London • New York

This translation first published by NLB, 1972;
copyright © NLB, 1972.
First published as 'Studien über Autorität und Familie'
by Librairie Félix Alcan, 1936;
copyright © Herbert Marcuse, 1936.
This edition published by Verso, 2008;
copyright © Verso 2008.

1 3 5 7 9 10 8 6 4 2

Verso
UK: 6 Meard Street, London W1F 0EG
US: 20 Jay Street, Suite 1010, Brooklyn, NY 11201
www.versobooks.com

Verso is the imprint of New Left Books

ISBN-13: 978-1-84467-209-7

British Library Cataloguing in Publication Data
A catalogue record for this book is available from the British Library

Library of Congress Cataloging-in-Publication Data
A catalog record for this book is available from the Library of Congress

Printed in the US

Translator's note: My thanks are due to Ben Fowkes for his careful checking of the translation and for his invaluable assistance in finding the English-language sources for the quotations.

The authority relationship, as understood in these analyses, assumes two essential elements in the mental attitude of he who is subject to authority: a certain measure of freedom (voluntariness: recognition and affirmation of the bearer of authority, which is not based purely on coercion) and conversely, submission, the tying of will (indeed of thought and reason) to the authoritative will of an Other. Thus in the authority relationship freedom and unfreedom, autonomy and heteronomy, are yoked in the same concept and united in the single person of he who is subject. The recognition of authority as a basic force of social praxis attacks the very roots of human freedom: it means (in a different sense in each case) the surrender of autonomy (of thought, will, action), the tying of the subject's reason and will to pre-establish contents, in such a way that these contents do not form the 'material' to be changed by the will of the individual but are taken over as they stand as the obligatory norms for his reason and will. Yet bourgeois philosophy put the autonomy of the person right at the centre of its theory: Kant's teachings on freedom are only the clearest and highest expression of a tendency which has been in operation since Luther's essay on the freedom of the Christian man.

The concept of authority thus leads back to the concept of freedom: it is the practical freedom of the individual, his social freedom and its absence, which is at stake. The union of internal autonomy and external heteronomy, the disintegration of freedom in the direction of its opposite is the decisive characteristic of the concept of freedom which has dominated bourgeois theory since the Reformation. Bourgeois theory has taken very great pains to justify these contradictions and antagonisms.

The individual cannot be simultaneously free and unfree, autonomous and heteronomous, unless the being of the person is conceived as divisible and belonging to various spheres. This is quite possible once one ceases to hypostatize the I as the 'substance'. But the decisive factor is the mode of this division. If it is undertaken dualistically, the world is split in half: two relatively self-enclosed spheres are set up and freedom and unfreedom as totalities divided between them in such a way that one sphere is wholly a realm of freedom and the other wholly a realm of unfreedom. Secondly, what is internal to the person is claimed as the realm of freedom: the person as member of the realm of Reason or of God (as 'Christian', as 'thing in itself', as intelligible being) is free. Meanwhile, the whole 'external world', the person as member of a natural realm or, as the case may be, of a world of concupiscence which has fallen away from God (as 'man', as 'appearance'), becomes a place of unfreedom. The Christian conception of man as 'created being' 'between' *natura naturata* and *natura naturans*, with the unalterable inheritance of the Fall, still remains the unshaken basis of the bourgeois concept of freedom in German Idealism.

But the realm of freedom and the realm of unfreedom are not simply contiguous with or superimposed on each other. They are founded together in a specific relation. For freedom – and we must hold fast to this astonishing phrase despite its paradoxical nature – is the condition of unfreedom. Only because and in so far as man is free can he be unfree; precisely because he is 'actually' (as a Christian, as a rational person) completely free must he 'unactually' (as a member of the 'external' world) be unfree. For the full freedom of man in the 'external' world as well would indeed simultaneously denote his complete liberation from God, his enslavement to the Devil. This thought reappears in a secularized form in Kant: man's freedom as a rational being can only be 'saved' if as a sensual being he is entirely abandoned to natural necessity. The Christian doctrine of freedom pushes the liberation of man back until it pre-dates his actual history, which then, as the history of his unfreedom, becomes an 'eternal' consequence of this liberation. In fact, strictly speaking there is no liberation of man in history according to this doctrine or, to put it more precisely, Christian doctrine has good reasons for viewing such a liberation as primarily something negative and evil, namely the partial liberation from God, the achievement of freedom to sin (as symbolized in the Fall).

As an 'internally' free being man is born into a social order which, while it may have been posited or permitted by God, by no means represents the realm in which the existence or non-existence of man is decided upon. Whatever the nature of this order may be, the inner freedom of man (his pure belief and his pure will, provided they remain pure) cannot be broken in it. 'The power of the temporal authority, whether it does right or wrong, cannot harm the soul.'[1]

This absolute inwardness of the person, the transcendent nature of Christian freedom *vis-à-vis* all worldly authority, must at the same time mean an 'internal' weakening and breaking of the authority relationship, however completely the individual may submit externally to the earthly power. For the free Christian knows that he is 'actually' raised above worldly law, that his essence and his being cannot be assailed by it and that his subordination to the worldly authorities is a 'free' act, which he does not 'owe' them. 'Here we see that all works and all things are free to a Christian through his faith. And yet because the others do not yet believe, the Christian bears and holds with them, *although he is not obliged to do these things.* He does this freely. . .'[2] This simultaneous recognition and transcendence of the whole system of earthly authorities announces a very important element in the Christian-bourgeois doctrine of freedom – its *anti-authoritarian tendency*. The social meaning of this doctrine of freedom is not simply that the individual should submit *in toto* to any earthly authority and thus affirm *in toto* the given system of authorities at any time. The Protestantism of Luther and Calvin which gave the Christian doctrine of freedom its decisive form for bourgeois society, is bound up with the emergence of a new, 'young' society which had first to conquer its right to exist in a bitter struggle against existing authorities. Faced with the universal bonds of traditionalist feudalism it absolutely required the liberation of the individual within the earthly order as well

---

1. Luther, *Treatise on Good Works* (1520), in *Selected Writings of Martin Luther*, vol. I, Philadelphia, 1967, p. 174.
2. Op. cit., p. 118 (my italics).

(the individual free subject of the economic sphere later essentially became the model of its concept of the individual) – it required the liberation of the territorial sovereign from the authority of an internationally centralized Church and a central imperial power. It further required the liberation of the 'conscience' from numerous religious and ethical norms in order to clear the way for the rise of the bourgeoisie. In all these directions an *anti-authoritarian attitude* was necessary: and this will find its expression in the writers we shall discuss.

However, this anti-authoritarian tendency is only the complement of an order which is directly tied to the functioning of as yet opaque relationships of authority. From the very outset the bourgeois concept of freedom left the way open for the recognition of certain metaphysical authorities and this recognition permits external unfreedom to be perpetuated within the human soul.

This point announced a fresh duality in the Protestant-bourgeois concept of freedom: an opposition between Reason and Faith, rational and irrational (in fact anti-rational) factors. As opposed to the rational, 'calculating' character of the Protestant-capitalist 'spirit' which is often all too strongly emphasized, its irrational features must be particularly pointed out. There lies an ultimate lack of order at the very root of this whole way of life, rationalized and calculated down to the last detail as an 'ideal type', this whole 'business' of private life, family and firm: the accounts do not, after all, add up – neither in the particular, nor in the general 'business'. The everyday self-torture of 'inner-worldly asceticism' for the sake of success and profit still ultimately has to experience these things, if they really occur, as unforeseeable good fortune. The individual is confronted again and again with the fear of loss: the reproduction of the whole society is only possible at the price of continual crises. The fact that the production and reproduction of life cannot be rationally mastered by this society constantly breaks through in the theological and philosophical reflections on its existence. The terrible hidden God of Calvinism is only one of the most severe forms of such a breakthrough: Luther's strong defence of the 'unfree will' is a similar case, as is the yawning gulf between the pure form of the universal law and the material for its fulfilment in Kant's ethic. The bourgeoisie fought its greatest battles under the banner of 'Reason' but it is precisely bourgeois society which totally deprives reason of its realization. The sector of nature controlled by man through rational methods is infinitely larger than in the Middle Ages; society's material process of production has in many instances been rationalized down to the last detail – but as a *whole* it remains 'irrational'. These antagonisms appear in the most varied forms in the ambivalence of bourgeois relationships of authority: they are rational, yet fortuitous, objective, yet anarchic, necessary, yet bad.

I   Luther and Calvin

Luther's pamphlet *The Freedom of a Christian* brought together
for the first time the elements which constitute the specifically
bourgeois concept of freedom and which became the ideological
basis for the specifically bourgeois articulation of authority:
freedom was assigned to the 'inner' sphere of the person, to the
'inner' man, and at the same time the 'outer' person was subjected
to the system of worldly powers; this system of earthly
authorities was transcended through private autonomy and
reason; person and work were separated (person and office) with
the resultant 'double morality'; actual unfreedom and inequality
were justified as a consequence of 'inner' freedom and equality.
Right at the start of the work[1] are those two theses which,
following on from St Paul, express the internally contradictory
nature of the Christian concept of freedom with a conscious
emphasis on this paradoxical antinomy: 'A Christian is free and
independent in every respect, a bondservant to none. A Christian
is a dutiful servant in every respect, owing a duty to everyone'
(p. 357). And the dissolution of the contradiction: the first
sentence deals with 'the spiritual man, his freedom and his
supreme righteousness', the second sentence refers to 'the outer
man': 'In as far as he is free, he requires to do nothing. In as far
as he is a servant he must do everything' (p. 369). That expresses
clearly and sharply the dualistic doctrine of the two realms, with
freedom entirely assigned to the one, and unfreedom entirely
assigned to the other.

The more specific determinations of internal freedom are all
given in a counter-attack on external freedom, as negations of a
merely external state of freedom: 'No outer thing . . .' can make

1. Luther, *The Freedom of a Christian* (1520), in *Reformation Writings of Martin Luther*, vol. I, London, 1952, pp. 357ff.

the free Christian 'free or religious', for his freedom and his 'servitude' are 'neither bodily nor outward'; none of the external things 'touches the soul, either to make it free or captive' (pp. 357–8). Nothing which is in the world and stems from the world can attack the 'soul' and its freedom; this terrible utterance, which already makes it possible entirely to deprecate 'outer' misery and to justify it 'transcendentally', persists as the basis of the Kantian doctrine of freedom; through it, actual unfreedom is subsumed into the concept of freedom. As a result, a peculiar (positive and negative) ambiguity enters into this concept of freedom: the man who is enclosed in his inner freedom has so much freedom over all outer things that he becomes free *from* them – he doesn't even *have* them any more, he has no control over them (p. 367). Man no longer *needs* things and 'works' – not because he already has them, or has control over them, but because in his self-sufficient inner freedom he doesn't need them at all. 'If such works are no longer a prerequisite, then assuredly all commandments and laws are like broken chains; and if his chains are broken, he is assuredly free' (p. 362). Internal freedom here really seems to be transformed into external freedom. But the realm of external freedom which opens up is, from the standpoint of 'spiritual' salvation as a whole, a realm of 'things indifferent': what man is free to do here, what can be done or not done, is in itself irrelevant to the salvation of his soul. 'But "free" is that in which I have choice, and may use or not, yet in such a way that it profit my brother and not me.'[2] The 'free' things in this realm can also be called the 'unnecessary' things: 'Things which are not necessary, but are left to our free choice by God, and which we keep or not.'[3] Freedom is a total release and independence, but a release and independence which can never be freely fulfilled or realized through a deed or work. For this freedom so far precedes every deed and every work that it is always already realized when man begins to act. His freedom can never be the result of an action;

2. Luther, *The First Lent Sermon at Wittenberg* (9 March 1522), in *Selected Writings*, vol. II, p. 238.
3. Luther, *The Third Lent Sermon at Wittenberg* (11 March 1522), in op. ci vol. II, p. 243.

the action can neither add to nor diminish his freedom. Earthly 'works' are not done to fulfil the person who requires this; the fulfilment must have occurred 'through faith before all works' ... 'works follow, once the commandments have been met' (p. 364).

But what sense is left in the earthly work of man if it always lags behind fulfilment? For the 'internal' man there is in fact no sense at all. Luther is quite clear on this point: 'Works are lifeless things, they can neither honour nor praise God . . .' (loc. cit). A sentence pregnant with consequences: it stands at the beginning of a development which ends with the total 'reification' and 'alienation' of the capitalist world. Luther here hit on the nodal points of the new bourgeois *Weltanschauung* with great accuracy: it is one of the origins of the modern concept of the subject as person. Straight after he has proclaimed that works are 'lifeless things' he continues: 'But here we seek him who is not done, as works are, but is an initiator and a master of work' (loc. cit). What is sought is the person (or that aspect of the person) who (or which) is not done (by another) but who is and stays the real subject of activity, the real master over his works: the autonomously acting person. And at the same time – this is the decisive point – this person is sought in contradistinction to his ('lifeless') works: as the negation and negativity of the works. Doer and deed person and work are torn asunder: the person as such essentially never enters into the work, can never be fulfilled in the work, eternally precedes any and every work. The true human subject is never the subject of *praxis*. Thereby the person is relieved to a previously unknown degree from the responsibility for his praxis, while at the same time he has become free for all types of praxis: the person secure in his inner freedom and fullness can only now really throw himself into outer praxis, for he knows that in so doing nothing can basically happen to him. And the separation of deed and doer, person and praxis, already posits the 'double morality' which, in the form of the separation of 'office' and 'person' forms one of the foundation stones of Luther's ethics:[4]

4. Luther, *Sermon on the Ban* (1520), in *Luther's Works*, vol. 39, ed. H. Lehmann,

later we shall have to return to the significance of this divorce.

But we have not yet answered the question. What meaning can the praxis of a person thus separated from his works still possess? His praxis is at first completely 'in vain': it is obvious that man as a person 'is free from all commandments, and quite voluntarily does all that he does without recompense, and apart from seeking his own advantage or salvation. He already has sufficient, and he is already saved through his faith and God's grace. What he does is done just to please God' (p. 372). The person does not need the works, but they must nevertheless be done, so that 'man may not go idle and may discipline and care for his body' (p. 371). The praxis which has been separated from the being of the person serves the sinful body, which is struggling against inner freedom, as a discipline, an incentive and a divine service. Here we cannot elaborate any further on this conception of inner-worldly ascetism, or its suitability for rationalizing life and its various modifications in Lutheranism and Calvinism; we need only point out that it is implanted in the Protestant concept of freedom, to which we now return.

Man is embedded in a system of earthly order which by no means corresponds to the fundamental teachings of Christianity. This contradiction provides a function for the 'double morality' as combined with the sharp distinction between the 'Christian' and the worldly human existence, between 'Christian' morality and 'external' morality, which is the motive force in offices and works'. The former refers only to the 'inner' man: his 'inner' freedom and equality,[5] his 'inner' poverty, love and happiness (at its clearest in Luther's interpretation of the Sermon on the Mount, 1530).[6] The 'external' order, on the other hand, is measured completely by the rules to which praxis and works are subjected when taken in isolation from the person. It is very

---

Philadelphia, 1970, p. 8; and *Whether Soldiers, Too, Can Be Saved* (1526), in *Selected Writings*, vol. III, p. 434.

5. Luther, *Temporal Authority: To What Extent It Should be Obeyed* (1523), *Selected Writings*, vol. II, p. 307: emphasizing the exclusively 'inner' equality of men.

6. Translated into English in *Luther's Works*, vol. 21, ed. J. Pelikan, Philadelphia, 1956, pp. 3ff.

characteristic that here – in accordance with the idea of praxis as the discipline and service performed by an utterly sinful existence – the earthly order appears essentially as a system of 'authorities' and 'offices', as an order of universal subordination, and that these authorities and offices in turn essentially appear under the sign of the 'sword'. (In one of his fiercest passages about worldly authority, still in anti-authoritarian idiom, Luther calls the Princes of God 'jailers', 'hangmen' and 'bailiffs'.)[7] This whole system of subordination to authorities and offices can admittedly be justi-fied as a whole by referring to the ordinances of God: it has been set up to punish the bad, to protect the faithful and to preserve the peace – but this justification is by no means sufficient to sanction the system of subordination that exists at any one time, the particular office or the particular authority and the way it uses the 'sword'. Can an unchristian authority be ordained by God and lay claim to unconditional subordination? Here the separation of office and person opens up a path which has far reaching conse-quences: it holds fast to the unconditional authority of the office, while it surrenders the officiating person to the fate of possible rejection. 'Firstly a distinction must be made: office and person, work and doer, are different things. For an office or a deed may well be good and right in itself which is yet evil and wrong if the person or doer is not good or right or does not do his work properly.'[8] There was already a separation of this kind before Luther, in Catholicism, but in the context of the doctrine of the inner freedom of the Christian man and of the rejection of any justification by 'works' it paves the way for the theoretical justification of the coming, specifically bourgeois, structure of authority.

The dignity of the office and the worthiness of the officiating person no longer coincide in principle. The office retains its unconditional authority, even if the officiating person does not deserve this authority. From the other side, as seen by those

---

7. *Selected Writings*, vol. II, p. 303.

8. *Selected Writings*, vol. III, p. 434, and cf. *Werke*, ed. Buchwald, Berlin 1905, vol. III, pt 2, p. 393.

subject to authority, in principle every 'under-person' is equal as a person to every 'over-person': with regard to 'inner' worthiness he can be vastly superior to the authority. Despite this he must give it his complete obedience. There is a positive and a negative justification for this. Negatively: because after all the power of the wordly authority only extends over 'life and property, and external affairs on earth',[9] and thus can never affect the being of the person, which is all that matters. Positively; because without the unconditional recognition of the ruling authorities the whole system of earthly order would fall apart, otherwise 'everyone would become a judge against the other, no power or authority, no law or order would remain in the world; there would be nothing but murder and bloodshed'.[10] For in this order there is no way in which one person can measure the worthiness of another or measure right and wrong at all. The system of authority proclaimed here is only tenable if earthly justice is taken out of the power of the people or if the existing injustice is included in the concept of earthly justice. God alone is judge over earthly injustice, and 'what is the justice of the world other than that everyone does what he owes in his estate, which is the law of his own estate: the law of man or woman, child, servant or maid in the house, the law of the citizen or of the city in the land . . .'.[11] There is no tribunal that could pass judgement on the existing earthly order – except its own existing tribunal: 'the fact that the authority is wicked and unjust does not excuse tumult and rebellion. For it is not everyone who is competent to punish wickedness, but only the worldly authority which wields the sword . . .'.[12] And just as the system of worldly authorities is its own judge in matters of justice, so also in matters of mercy: the man who appeals to God's mercy in the face of the blood and terror of this system is turned away. 'Mercy is neither here nor there; we are now speaking of the word of God, whose will is that

9. *Temporal Authority* (1523), in *Selected Writings*, vol. II, p. 295.
10. *Admonition to Peace: A Reply to the Twelve Articles of the Peasants in Swabia* (1525), in *Selected Writings*, vol. III, p. 327.
11. *Werke*, ed. Buchwald, Berlin, 1905, vol. III, pt 2, p. 300.
12. *Admonition to Peace*, in *Selected Writings*, vol. III, p. 325.

the King be honoured and rebels ruined, and who is yet surely as merciful as we are.' 'If you desire mercy, do not become mixed up with rebels, but fear authority and do good.'[13]

We are looking here only at those consequences which arise from this conception for the new social structure of authority. A rational justification of the existing system of worldly authorities becomes impossible, given the absolutely transcendental character of 'actual' justice in relation to the worldly order on the one hand, and the separation of office and person and the essential immanence of injustice in earthly justice on the other. In the Middle Ages authority was tied to the particular bearer of authority at the time; it is the 'characteristic of him who communicates the cognition of a judgement'[14] and as a 'characteristic' it is inseparable from him; he always 'has' it for particular reasons (which again can be rational or irrational). Now the two are torn apart: the particular authority of a particular worldly bearer of authority can now only be justified if we have recourse to authority in general. Authority must exist, for otherwise the worldly order would collapse. The separation of office and person is only an expression for the autonomization (*Verselbständigung*) and reification of authority freed from its bearer. The authority-system of the existing order assumes the form of a set of relationships freed from the actual social relationships of which it is a function; it becomes eternal, ordained by God, a second 'nature' against which there is no appeal. 'When we are born God dresses and adorns us as another person, he makes you a child, me a father, the one a lord, the other a servant, this one a prince, that one a citizen and so on.'[15] And Luther accuses the peasants who protested against serfdom of turning Christian freedom into 'something completely of the flesh': 'Did not Abraham and other patriarchs and prophets also have slaves?'[16]

13. *An Open Letter on the Harsh Book against the Peasants* (1525), in *Selected Writings*, vol. III, p. 371.
14. Grimmich, *Lehrbuch der theoretischen Philosophie auf thomistischer Grundlage*, Freiburg, 1893, p. 177.
15. *Werke*, ed. Buchwald, Berlin, 1905, vol. II, pt 2, p. 296.
16. *Admonition to Peace*, in *Selected Writings*, vol. III, p. 339.

It is no coincidence that it is the essence of 'Christian freedom' which is held up to the rebellious peasants, and that this does not make them free but actually confirms their slavery. The recognition of actual unfreedom (particularly the unfreedom caused by property relations) is in fact part of the sense of this concept of freedom. For if 'outer' unfreedom can attack the actual being of the person, then the freedom or unfreedom of man is decided on earth itself, in social praxis, and man is, in the most dangerous sense of the word, free from God and can freely become himself. The 'inner', *a priori* freedom makes man completely helpless, while seeming to elevate him to the highest honour: it logically precedes all his action and thought, but he can never catch his freedom up and take possession of it.

In the young Marx's formulation, this unfreedom conditioned by the internalization of freedom, this dialectic between the release from old authorities and the establishment of new ones is a decisive characteristic of Protestantism: 'Luther, without question, defeated servitude through devotion, but only by substituting servitude through conviction. He shattered the faith in authority, by restoring the authority of faith. . . . He freed man from external religiosity by making religiosity the innermost essence of man.'[17]

One of the most characteristic passages for the unconditional acceptance of actual unfreedom is Luther's admonition to the

17. Marx, *Introduction to a Contribution to the Critique of Hegel's Philosophy of Right*, in *Karl Marx: Early Writings*, trans. T. B. Bottomore, London, 1963, p. 53.

The contradiction between anti-authoritarian and authoritarian tendencies which pervades the whole of Luther's work has been clearly elaborated by R. Pascal, *The Social Basis of the German Reformation*, London, 1933. Pascal shows that this contradiction is determined by the social and economic situation of the urban petty bourgeoisie, to whose interests Luther's Reformation corresponds. Pascal further strongly emphasizes the basically authoritarian character of Lutheranism, into which the anti-authoritarian streams are ultimately also fitted, so that after the achievement of the socially necessary economic and psychological liberations they work completely in the interests of the stabilization and strengthening of the existing world order. Even on the rare occasions when Luther breaks his doctrine of unconditional obedience to the worldly authority (as in 1531 with regard to the question of armed resistance to the Emperor by the Princes, after Luther had finally had to abandon his hope of winning the Emperor for the Protestant cause), the position he takes is by no means revolutionary but conservative: the Emperor appears as the wanton destroyer of an order which must be preserved under all circumstances.

Christian slaves who had fallen into the hands of the Turks, telling them not to run away from their new lords or to harm them in any other way: 'You must bear in mind that you have lost your freedom and become someone's property, and that without the will and knowledge of your master you cannot get out of this without sin and disobedience.' And then the interesting justification: 'For thus you would rob and steal your body from your master, which he has bought or otherwise acquired, after which it is not your property but his, like a beast or other goods in his possession.'[18] Here, therefore, certain worldly property and power relationships are made the justification of a state of unfreedom in which even the total abandonment of the Christian to the unbeliever is of subordinate importance to the preservation of these property relationships.[19]

With the emergence of the independence of worldly authority, and its reifications, the breach of this authority, rebellion and disobedience, becomes the social sin pure and simple, a 'greater sin than murder, unchastity, theft, dishonesty and all that goes with them.'[20] 'No evil deed on earth' is equal to rebellion; it is a 'flood of all wickedness'.[21] The justification which Luther gives for such a hysterical condemnation of rebellion reveals one of the central features of the social mechanism. While all other evil deeds only attack individual 'pieces' of the whole, rebellion attacks 'the head itself'. The robber and murderer leave the head that can punish them intact and thus give punishment its chance; but rebellion

18. *On War Against the Turk* (1529), in *Selected Writings*, vol. IV, p. 42.

19. Thomas Münzer's attack on Luther deals precisely with this connection between Luther's concept of authority and a particular property order: 'The poor flatterer wants to cover himself with Christ in apparent goodness. . . . But he says in his book on trading that one can with certainty count the princes among the thieves and robbers. But at the same time he conceals the real origin of all robbery. . . . For see, our lords and princes are the basis of all profiteering, theft and robbery; they make all creatures their property. The fish in the water, the birds of the air, the plants on the earth must all be theirs (Isaiah 5). Concerning this they spread God's commandment among the poor and say that God has commanded that you shall not steal, but it does them no good. So they turn the poor peasant, the artisan and all living things into exploiters and evil-doers' (*Hoch verursachte Schutzrede* (1525), in *Flugschriften aus der Reformationszeit*, vol. X, Halle, 1893, p. 25).

20. *Treatise on Good Works*, in *Selected Writings*, vol. I, p. 163.

21. *An Open Letter*, in *Selected Writings*, vol. III, p. 381.

'attacks punishment itself' and thereby not just disparate portions of the existing order, but this order itself (op. cit., pp. 380–81), which basically rests on the credibility of its power of punishment and on the recognition of its authority. 'The donkey needs to feel the whip and the people need to be ruled with force; God knew that well. Hence he put a sword in the hands of the authorities and not a featherduster' (op. cit., p. 376). The condition of absolute isolation and atomization into which the individual is thrown after the dissolution of the medieval universe appears here, at the inception of the new bourgeois order, in the terribly truthful image of the isolation of the prisoner in his cell: 'For God has fully ordained that the under-person shall be alone unto himself and has taken the sword from him and put him into prison. If he rebels against this and combines with others and breaks out and takes the sword, then before God he deserves condemnation and death.'[22]

Every metaphysical interpretation of the earthly order embodies a very significant tendency: a tendency towards *formalization*. When the existing order, in the particular manner of its materiality, the material production and reproduction of life, becomes ultimately valueless with regard to its 'actual' fulfilment, then it is no more than the form of a social organization as such, which is central to the organization of this life. This form of a social order ordained by God for the sinful world was for Luther basically a system of 'over-persons' and 'under-persons'. Its formalization expressed itself in the separation of dignity and worthiness, of office and person, without this contradiction giving any rightful basis for criticism or even for the reform of this order. It was thus that the encompassing system of worldly authorities was safeguarded: it required unconditional obedience (or, if it intruded on 'Christian freedom', it was to be countered with spiritual weapons or evaded).

But danger threatened from another quarter. Initially, the unconditional freedom of the 'person', proclaimed by Luther, encouraged an anti-authoritarian tendency, and, indeed, precisely

22. *Selected Writings*, vol. III, p. 466.

on account of the reification of authority. The dignity of the office was independent of the worthiness of its incumbent; the bourgeois individual was 'privately' independent of authority. The assertion of Christian freedom and the allied conception of a 'natural realm' of love, equality and justice was even more destructive. Although it was separated from the existing social order by an abyss of meaning, it must still have threatened the completely formalized social order simply by its claims and its full materiality. The ideas of love, equality and justice, which were still effective enough even in their suppressed Lutheran form, were a recurrent source of anxiety to the rising bourgeois society owing to their revolutionary application in peasant revolts, Anabaptism and other religious sects. The smoothing-out of the contradictions appearing here, and the incorporation of these destructive tendencies into the bourgeois order, was one of the major achievements of Calvin. It is significant that this synthesis was only possible because the contradictions were simultaneously breaking out anew in a different dimension – although now in a sphere no longer transcending the bourgeois order as a whole but immanent in it. The most important marks of this tendency are Calvin's 'legalism' and his doctrine of the 'right to resist'.

It has often been pointed out in the relevant literature that in Calvin the Lutheran 'natural law' disappears. The dualism of the two 'realms' is removed:[23] admittedly Calvin too had sharply to emphasize that (precisely because of his increased interest in the bourgeois order) 'the spiritual kingdom of Christ and civil government are things very widely separated'[24] but the Christian realm of freedom is no longer effective as the material antithesis of the earthly order. In the face of the completely sinful and evil world there is ultimately only the person of God who, through the sole mediation of Christ, has chosen individuals for redemption

23. Beyerhaus, *Studien zur Staatsanschauung Calvins*, Berlin, 1910, points out that although 'theoretically' a distinction is made between the two spheres, 'practically' they become a unity precisely in the realization of Calvin's idea of the state (p. 50).

24. Calvin, *Institutes of the Christian Religion*, trans. F. L. Battles, London, 1961, Book IV, ch. XX, para. 1.

by a completely irrational system of predestination. Luther had been greatly disturbed by the tensions between his teaching and the teachings of the 'Sermon on the Mount', where the transcendence of the existing order is most clearly expressed and a devastating critique of this order made, which no degree of 'internalization' could ever completely suppress: in Calvin these tensions no longer exist. The more inexorably Calvin elaborates the doctrine of eternal damnation, the more the positive biblical promises lose their radical impulse.[25] The way is made clear for a view of the wordly order which does not recognize its dubious antithesis. This does not mean that the world is somehow 'sanctified' in the Christian sense: it is and remains an order of evil men for evil men, an order of concupiscence. But in it, as the absolutely prescribed and sole field for their probation, Christians must live their life to the honour and glory of the divine majesty, and in it the success of their praxis is the *ratio cognoscendi* (reason of knowing) of their selection. The *ratio essendi* (reason of existence) of this selection belongs to God and is eternally hidden from men. Not love and justice but the terrible majesty of God was at work in the creation of this world, and the desires and drives, the hopes and laments of men are correspondingly directed not towards love and justice but towards unconditional obedience and humble adoration. Very characteristically, Calvin conceived original sin, i.e. the act which once and for all determined the being and essence of historical man, as disobedience, *inoboedientia*,[26] or as the crime of lese-majesty (while in St Augustine's interpretation of original sin as *superbia* [overwhelming pride] – which Calvin aimed to follow here – there is still an element of the defiant freedom of the self-affirming man). And obedience is also the mechanism which holds the wordly order together: a system, emanating from the family, of *subjectio* and *superioritas*, to which God has given his name for protection: 'The titles of Father, God and Lord, all meet in him alone, and hence, whenever any one of

25. H. Engelland, *Gott und Mensch bei Calvin*, Munich, 1934, pp. 113ff.

26. Cf. Barnikel, *Die Lehre Calvins vom unfreien Willen . . .*, Bonn dissertation, 1927, pp. 104ff; Beyerhaus, op. cit., p. 79.

them is mentioned our mind should be impressed with the same feeling of reverence' (*Institutes*, Book II, ch. VIII, para. 35).

By freeing the worldly order from the counter-image of a Christian realm of love, equality and justice and making it as a whole a means for the glorification of God, the formalization operative in Luther is withdrawn; the sanction granted it now also affects its materiality: '. . . in all our cares, toils, annoyances, and other burdens, it will be no small alleviation to know that all these are under the superintendence of God. The magistrate will more willingly perform his office, and the father of the family confine himself to his proper sphere. Every one in his particular mode of life will, without refining, suffer its inconveniences, cares, uneasiness, and anxiety, persuaded that God has laid on the burden' (op. cit., Book III, ch. X, para. 6). The new direction manifests itself in the often described activism and realism of Calvin's disciples: in the concept of an occupation as a vocation, in Calvin's 'state rationalism', in his extensive and intensive practico-social organization. With the abolition of Luther's formalization, the separation of office and person and the 'double morality' linked with it also disappear in Calvin (although it will be shown that this does not remove the reification of authority, i.e. the understanding of it as an element of a natural or divine feature of an institution or a person instead of as a function of social relationships): the religious moral law – and essentially in the form represented in the decalogue, which it is claimed is also a 'natural' law – is regarded as the obligatory norm for the practical social organization of the Christian 'community'. This was a step of great significance. It is true that the decalogue complied to a much greater degree with the demands of the existing social order than with the radical transcendental Christianity of the New Testament, and that it provided a considerably greater amount of latitude. Nevertheless, the new form of the law stabilized a norm, against which the officiating authorities could be 'critically' measured. 'But now the whole doctrine is pervaded by a spirit which desires to see society shaped and moulded for a definite purpose, and a spirit which can criticize law and authority according to the eternal standards of divine and natural law.'[27] Luther's irrationalist doctrine of authority as 'power for power's sake', as Troeltsch characterized it in a much disputed phrase, has been abandoned. In so far as obedience to the officiating authority leads to a transgression of the law, this authority loses its right to obedience.[28] It is a straight line from here to the struggle of the *Monarchomachi* against absolutism. From a source very close to Calvin, from his pupil, Théodore de Beza, comes the famous work *De jure magistratum in subditos* which presents the opinion that 'even armed revolution is permissible, if no other means remain . . .'.[29]

Yet these tendencies already belong to the later development of the bourgeoisie; in Calvin the right to resist in the face of worldly authorities is in principle limited from the start. Immediately after his warning to unworthy princes ('May the princes hear, and be afraid') Calvin continues: 'But let us at the same time guard most carefully against spurning or violating the venerable and majestic authority of rulers, an authority which God has sanctioned by the surest edicts, although those invested with it should be most unworthy of it, and, as far as in them lies, pollute it with their iniquity. Although the lord takes vengeance on unbridled domination, let us not therefore suppose that that vengeance is committed to us, to whom no command has been given but to obey and suffer. I speak only of private men' (*Institutes*, Book IV, ch. XX, para. 31). Worldly authority retains its independence and its reification. And in a characteristic modification of the Lutheran concept of the *homo privatus* as a free person, this *homo privatus* is now primarily unfree: he is the man who obeys and suffers. In no case is the *homo privatus* entitled to change the system of officiating authorities:[30] 'The subject as a private person has no independent political rights, rather he has the ethical-religious

27. Troeltsch, *The Social Teaching of the Christian Churches*, trans. O. Wyon, vol. II, London, 1931, p. 616.

28. Ibid., p. 618.

29. Ibid., p. 629.

30. Troeltsch, op. cit., p. 616; Lobstein, *Die Ethik Calvins*, Strasburg, 1877, p. 116.

duty to bear patiently even the extremities of oppression and persecution.'[31] Even in the case of the most blatant transgression of the Law, when obedience to the worldly authority must lead to disobedience to God, Calvin allows only a 'right of passive resistance'. Where the Christian organization of society is actually already under attack the right of veto is allowed only to the lower magistrates themselves, never to the 'people' or to any postulated representatives of the people. And so in Calvin too we encounter the Lutheran idea of the immanence of the law within the existing system of worldly authorities: decisions regarding their rightness or wrongness are made exclusively within their own order, among themselves.

The direct ordination of the system of worldly authorities by God, when combined with the Calvinist concept of God as the absolute 'sovereign', means both a strengthening and a weakening of worldly authorities – one of the many contradictions which arose when the Christian idea of transcendence ceased to be effective. Direct divine sanction increases the power of the earthly authorities: 'The lord has not only declared that he approves of and is pleased with the function of magistrates, but also strongly recommended it to us by the very honourable titles which he has conferred upon it',[32] – although at the same time it should not thereby under any circumstances be allowed to lead to a diminution or a division of the sovereignty of God. All worldly power can only be a 'derivative right': authority is a 'jurisdiction as it were delegated by God'. But for the people this delegacy is irremovable and irrevocable.[33] The relationship of God to the world appears essentially as the relationship of an unlimited sovereign to his subjects. Beyerhans has pointed out, with due caution, although clearly enough, that Calvin's concept of God 'betrays the influence of worldly conceptions of law and power'.[34]

A good index for the status of Protestant-bourgeois man in relation to the system of worldly order is the contemporary version

31. Beyerhaus, op. cit., p. 97.
32. *Institutes*, Book IV, ch. XX, para. 4.
33. Beyerhaus, op. cit., p. 87.
34. Beyerhaus, op. cit., p. 79.

of the concept of freedom. On the road from Luther to Calvin the concept of *libertas christiana* becomes a 'negative' concept. 'Christian freedom . . . is not understood positively as mastery over the world but in a purely negative manner as the freedom from the damning effect of the law.'[35] Calvin's interpretation of *libertas christiana* was essentially based on the polemic interpretation of Christian freedom. Luther's concept of freedom had not been positive in Lobstein's sense either. But in the establishment of an unconditional 'inner' freedom of the person there was none the less an element which pointed forward towards the real autonomy of the individual. In Calvin this moment is forced into the background. The threefold definition of *libertas christiana* in the *Institutes* (Book III, ch. XIX, paras 2, 4, 7) is primarily negative in all its three elements: (*a*) freedom of the conscience from the necessity of the law – not indeed as a higher authority to be relied on against the validity of the law, but (*b*) as 'voluntary' subordination to the law as to the will of God: 'they voluntarily obey the will of God, being free from the yoke of the law itself',[36] and (in the sense already indicated in Luther) (*c*) freedom from external things 'which in themselves are but matters indifferent', and which 'we are now at full liberty either to use or omit'.[37] We should stress, precisely in view of this last definition that, combined with Calvin's idea of vocation and of probation in the vocation, the adiaphorous character of the external things has become a strong ideological support for Protestant economic praxis under capitalism. The negativity of this concept of freedom is revealed here by its inner connection with a social order which despite all external rationalization is basically anti-rational and anarchic, and which, in view of its final goal, is itself negative.

What remains as a positive definition of freedom is freedom in the sense of freedom to obey. For Calvin it is no longer a problem that 'spiritual freedom can very well coexist with political servitude' (*Institutes*, Book IV, ch. XIX, para 1). But the difficulty of

35. Lobstein, op. cit., p. 148.
36. Op. cit., Book III, ch. XIX, para. 4. Cf. in I Peter, ch. 2, verse 16: 'The purpose of our liberty is this, that we should obey more readily and more easily' (Lobstein, op. cit., p. 37).   37. Op. cit., Book III, ch. XIX, para. 7.

uniting freedom and unfreedom reappears in the derivative form of the union of freedom and the unfree will. Calvin agrees with Luther that Christian freedom not only does not require free will, but that it excludes it. Both Luther and Calvin base the unfree will on a power which man simply cannot eradicate: on the depravity of human nature which arose from the Fall and the absolute omnipotence of the divine will. The unfree will is an expression of the eternal earthly servitude of men:[38] it cannot and may not be removed without exploding the whole Christian-Protestant conception of man and the world. For Calvin, not only man's sensuality but also his reason is ultimately corrupt. This provides the theological justification for an anti-rationalism which strongly contrasts with Catholic teaching. In the Catholic doctrine there was still an awareness that reason and freedom are correlative concepts, that man's rationality will be destroyed if it is separated from the free possibility of rational acting and thinking. For Thomas Aquinas, man, as a rational animal, is necessarily also free and equipped with free will: 'And forasmuch as man is rational is it necessary that man have a free will.'[39] In Luther reason itself attests to the fact 'that there is no free will either in man or in any other creature'.[40] Reason is here characteristically appraised as the index of human unfreedom and heteronomy: thus we read in Luther's *Treatise on Good Works*, after the interpretation of the first four commandments: 'These four preceding commandments do their work in the mind, that is, they take man prisoner, rule him and bring him into subjection so that he does not rule himself, does not think himself good, but rather acknowledges his humility and lets himself be led, so that his pride is restrained.'[41] To this should be added the loud warnings which Luther gives against an overestimation of human reason and its realm ('We must not start something by trusting in

---

38. 'For where there is servitude, there is also necessity.' Cf. Barnikel, op. cit., p. 113.

39. *Summa Theol.* I, quaestio 83, art. 1.

40. *Martin Luther on the Bondage of the Will*, translation of *De Servo Arbitrio*, (1525) by J. I. Packer and O. R. Johnston, London, 1957, p. 317; cf. Barnikel, op. cit., p. 46. 41. *Treatise on Good Works*, in *Selected Writings*, vol. I, p. 182.

the great power of human reason . . . for God cannot and will not suffer that a good work begin by relying upon one's own power and reason'),[42] and the rejection of a rational reform of the social order in Calvin. This is all a necessary support for the demand for unconditional subordination to independent and reified wordly authorities, for which any rational justification is rejected.

But this doctrine of the 'unfree will' contains a new contradiction which must be resolved. How can man conceivably still be responsible for himself if the human will is fully determined? Man's responsibility must be salvaged: the Christian doctrine of sin and guilt, the punishment and redemption of man requires it, but the existing system of worldly order requires it too, for – as we have indicated – this system for both Luther and Calvin is essentially tied to the mechanism of guilt and punishment. Here the concept of 'psychological freedom' offers a way out: Calvin expounds the concept of a necessity (*necessitas*) which is not coercion (*coactio*) but a 'spontaneous necessity'. The human will is necessarily corrupt and necessarily chooses evil. This does not mean, however, that man is forced, 'against his will' to choose evil; his enslavement in sin is a 'voluntary enslavement' (*servitus voluntaria*). 'For we did not consider it necessary to sin, other than through weakness of the will; whence it follows that this was voluntary.'[43] Thus despite the *necessitas* of the will, responsibility can be ascribed for human deeds. The concept of enslavement or voluntary necessity signifies one of the most important steps forward in the effort to perpetuate unfreedom in the essence of human freedom: it remains operative right up until German Idealism. Necessity loses its character both as affliction and as the removal of affliction; it is taken from the field of man's social praxis and transferred back into his 'nature'. In fact necessity is restored to nature in general and thus all possibility of overcoming it is removed. Man is directed not towards increasingly overcoming necessity but towards voluntarily accepting it.

---

42. *To the Christian Nobility of the German Nation Concerning the Reform of the Christian Estate* (1520), in *Selected Writings*, vol. I, p. 261.
43. Calvin, *Opera*, vol. VI, p. 280.

As is well known, a programmatic reorganization of the family and a notable strengthening of the authority of the *pater familias* took place in the context of the bourgeois-Protestant teachings of the Reformation. It was firstly a necessary consequence of the toppling of the Catholic hierarchy; with the collapse of the (personal and instrumental) mediations it had set up between the individual and God, the responsibility for the salvation of the souls of those not yet responsible for themselves, and for their preparation for the Christian life, fell back on the family and on its head, who was given an almost priestly consecration. On the other hand, since the authority of the temporal rulers was tied directly to the authority of the *pater familias* (all temporal rulers, all 'lords' become 'fathers'), their authority was consolidated in a very particular direction. The subordination of the individual to the temporal ruler appears just as 'natural', obvious, and 'eternal' as subordination to the authority of the father is meant to be, both deriving from the same divinely ordained source. Max Weber emphasizes the entry of 'calculation into traditional organizations brotherhood' as a decisive feature of the transformation of the family through the penetration of the 'capitalist spirit': the old relationships of piety decay as soon as things are no longer shared communally within the family but 'settled' along business lines.[44] But the obverse side of this development is that the primitive, 'naïve' authority of the *pater familias* becomes more and more a planned authority, which is artificially generated and maintained.

The key passages for the doctrine of the authority of the *pater familias* and of the 'derivation' of worldly authorities from it are Luther's exegeses of the Fourth Commandment in the *Sermon on Good Works* and in the *Large Catechism*, and Calvin's interpretation in the *Institutes*, Book II, ch. VIII. Luther directly includes within the Fourth Commandment 'obedience to over-persons, who have to give orders and rule', although there is no explicit mention of these. His justification, thus, characteristically, runs

44. Max Weber, *General Economic History*, trans. F. H Knight, Glencoe, 1930, p. 356.

as follows: 'For all authority has its root and source in parental authority. For where a father is unable to bring up his child alone, he takes a teacher to teach him; if he is too weak, he takes his friend or neighbour to help him; when he departs this life, he gives authority to others who are chosen for the purpose. So he must also have servants, men and maids, under him for the household, so that all who are called master stand in the place of parents, and must obtain from them authority and power to command. Wherefore in the bible they are all called fathers.'[45] Luther saw clearly that the system of temporal authorities constantly depends on the effectiveness of authority within the family. Where obedience to father and mother are not in force 'there are no good ways and no good governance. For where obedience is not maintained in houses, one will never achieve good governance, in a whole city, province, principality or kingdom'.[46] Luther saw that the system of society which he envisaged depended for its survival as such on the continued functioning of parental authority; 'where the rule of the parents is absent, this would mean the end of the whole world, for without governance it cannot survive'.[47] For the maintenance of this world 'there is no greater dominion on earth than the dominion of the parents',[48] for there is 'nothing more essential than that we should raise people who will come after us and govern'.[49] The wordly order always remains in view as a system of rulers and ruled to be maintained unquestioningly.

On the other hand, however, parental authority (which is always paternal authority in Luther) is also dependent on worldly authority: the *pater familias* is not in a position to carry out the upbringing and education of the child on his own. Alongside the parents, there is the school, and the task of educating the future rulers in all spheres of social life is impressed on it too. Luther

45. *The Large Catechism* (1529), in *Luther's Primary Works*, trans. H. Wace and C. A. Buchheim, London, 1896, p. 58.
46. Quoted from *Luther als Pädagog*, ed. E. Wagner (Klassiker der Pädagogik, Vol. II), Langensalza, 1892, p. 70.
47. Ibid., p. 73.
48. Ibid., p. 64.
49. Ibid., p. 119.

sees the reason for divinely sanctioned parental authority in the breaking and humiliation of the child's will: 'The commandment gives parents a position of honour so that the self-will of the children can be broken, and they are made humble and meek':[50] 'for everyone must be ruled, and subject to other men'.[51] Once again it is the image of the wordly order as universal subordination and servitude which is envisaged by Luther, a servitude whose simple 'must' is no longer even questioned. The freedom of the Christian is darkened by the shadow of the coming bourgeois society; the dependence and exploitation of the greatest part of humanity appears implanted in the 'natural' and divine soil of the family; the reality of class antagonisms is turned into the appearance of a natural-divine hierarchy, exploitation becomes the grateful return of gifts already received. For that is the second ground for unconditional obedience: 'God gives to us and preserves to us through them [the authorities] as through our parents, our food, our homes, protection and security';[52] 'we owe it to the world to be grateful for the kindness and benefits that we have received from our parents.'[53] And servants and maids ought even to 'give up wages' out of pure gratefulness and joy at being able to fulfil God's commandment in servitude.[54]

The personal characteristics which the coming social order wishes to produce require a change in all human values from earliest childhood. Honour (*Ehrung*) and fear (*Furcht*) or, taken together, reverence (*Ehrfurcht*) take the place of love as the determining factor in the relationship between the child and its parents.[55] 'For it is a far higher thing to honour than to love, since honouring does not simply comprise love [but] obedience, humility and reverence, as though towards some sovereign hidden there.'[56] The terrible majesty of Calvin's God comes to the surface

50. *Selected Writings*, vol. I, p. 168.
51. Op. cit., p. 164.
52. *Luther's Primary Works*, p. 60.
53. Op. cit., p. 56.
54. Op. cit., p. 59.
55. For a contrary passage, cf. *Luther als Pädagog*, p. 64.
56. *Luther's Primary Works*, p. 52.

in the authority of the *pater familias*. It is precisely discipline and fear which raises honouring one's parents above love: 'honour is higher than mere love, for it includes within it a kind of fear which, combined with love, has such an effect on a man that he is more afraid of injuring them than of the ensuing punishment'.[57] Just as disobedience is the greatest sin, obedience is the highest 'work' after those commanded in Moses's first tablet; 'so that to give alms and all other work for one's neighbour is not equal to this'.[58]

There are also passages in Luther in which parental and governmental authority are explicitly contrasted. Thus in the *Table Talks*: 'Parents look after their children much more and are more diligent in their care of them than the government is with its subjects. . . . The power of the father and mother is a natural and voluntary power and a dominion over children which has grown of itself. But the rule of the government is forced, an artificial rule.'[59] There is also some wavering on the question of the extension of the 'double morality' of office and person to parental authority. In the *Sermon on Good Works* (1520) Luther says: 'Where the parents are foolish and raise their children in a wordly manner, the children should in no way be obedient to them. For according to the first three Commandments God is to be held in higher esteem than parents.'[60] Nine years later, in the *Large Catechism*, he writes: 'Their [the parents'] condition or defect does not deprive them of their due honour. We must not regard their persons as they are, but the will of God, who ordered and arranged things thus.'[61]

In the passages quoted above one can see the tendency towards a separation of natural and social authority. Luther did not advance any further along the road from the 'natural' unity of the family to the 'artificial' and 'forced' unity of society; he was satisfied with establishing that the family is the 'first rule, in which

57. *Selected Writings*, vol. I, p. 163.
58. *Luther's Primary Works*, p. 56.
59. *Luther als Pädagog*, p. 53.
60. *Selected Writings*, vol. I, p. 166.
61. *Luther's Primary Works*, p. 52.

all other types of rule and domination have their origins'.[62] Calvin went a little further in this direction; he presents an exceptionally interesting psychological interpretation: 'But as this command to submit is very repugnant to the perversity of the human mind (which, puffed up with ambitious longings, will scarcely allow itself to be subjected) that superiority which is most attractive and least invidious is set forth as an example calculated to soften and bend our minds to the habits of submission. From that subjection which is most tolerable, the lord gradually accustoms us to every kind of legitimate subjection, the same principle regulating all.'[63]

Calvin agrees with Luther on the close association between subjection to authority in general and parental authority;[64] we saw how for him too the titles *Dominus* and *Pater* are interchangeable. But Calvin ascribes to the authority relationship of the family a quite definite function within the mechanism of subjection to social authorities. This function is psychological. Since subjection is actually repugnant to human nature, man should, through a type of subordination which by its nature is pleasant and will arouse the minimum of ill will, be gradually prepared for types of subordination which are harder to bear. This preparation occurs in the manner of a softening, bowing and bending; it is a continual habituation, through which man becomes accustomed to subjection. Nothing need be added to these words: the social function of the family in the bourgeois authority-system has rarely been more clearly expressed.

62. *Luther als Pädagog*, p. 70. Cf. Levin Schücking, *Die Familie im Puritanismus*, Leipzig, 1929, p. 89.
63. *Institutes*, Book II, ch. VIII, para. 35.
64. Troeltsch, op. cit., p. 603.

There are two ways of coming to an appreciation of the level
reached by Kant in dealing with the problem of authority: the
impact and the transformation of the 'Protestant ethic' could be
traced in the Kantian doctrine of freedom, or the problem of
authority and freedom could be developed immanently from the
centre of Kant's ethics. The inner connections between Lutheran
and Kantian ethics are plainly apparent. We shall point only to
the parallels given by Delekat:[1] the conception of 'inner' freedom
as the freedom of the autonomous person: the transfer of ethical
'value' from the legality of the 'works' to the morality of the
person; the 'formalization' of ethics; the centring of morality on
reverential obedience to duty as the secularization of 'Christian
obedience'; the doctrine of the actual unconditional authority of
worldly government. But with this method those levels of Kantian
ethics which cannot be comprehended under the heading of the
'Protestant ethic' would be given too short a shrift and appear in a
false light. The second way would indeed be a genuine approach,
but would require an extensive elaboration of the whole con-
ceptual apparatus of Kantian ethics, which we cannot provide
within the framework of this investigation. We will necessarily
have to choose a less adequate route: there are as it were two
central points around which the problematic of authority and
freedom in Kant's philosophy is concentrated: firstly, the philosophical foundation itself, under the heading of the autonomy of
the free person under the law of duty, and secondly the sphere of
the 'application' of ethics, under the heading of the 'right of
resistance'. In this second section Kant deals with the problem in
the context of a comprehensive philosophical interpretation of the

1. *Handbuch der Pädagogik*, ed. Nohl-Pallat, vol. I, Langensalza, 1928, pp.
221ff.

legal framework of bourgeois society.[2] The level of concreteness of the present treatment admittedly cannot compensate for its vast distance from the actual philosophical foundation, but it offers a good starting point.

In the small treatise, *Reply to the Question: What is Enlightenment?* (1784), Kant explicitly poses the question of the relation between social authority and freedom. To think and to act according to an authority is for Kant characteristic of 'immaturity', a 'self-inflicted immaturity', for which the person is himself to blame. This self-enslavement of man to authority has in turn a particular social purpose, in that civil society 'requires a certain mechanism, for some affairs which are in the interests of the community, whereby some members of the community must behave purely passively, so that they may, by means of an artificial consensus, be employed by the government for public ends (or at least deterred from vitiating them)'.[3] Bourgeois society has an 'interest' in 'disciplining' men by handling them in an authoritarian manner, for here its whole survival is at stake. In the closing note of his *Anthropology*, Kant described religion as a means of introducing such a discipline and as a 'requirement' of the constituted bourgeois order 'so that what cannot be achieved through external compulsion can be effected through the inner compulsion of the conscience. Man's moral disposition is utilized for political ends by the legislators. . . .'[4]

How can one square man's 'natural' freedom with society's interest in discipline? For Kant firmly believes that the free autonomy of man is the supreme law. It presupposes the exit of man from the state of immaturity which is his own fault'; this process is, precisely, 'enlightenment'. Nothing is needed for this

2. *Translator's note:* 'Bourgeois society' is here a translation of 'bürgerliche Gesellschaft', more usually rendered as 'civil society'. While Kant and Hegel certainly used the term in the sense of 'civil society', Marcuse used it in 1936 in the sense of 'bourgeois society', since, as he states in relation to Kant's concept of 'civil society', the 'actual features of bourgeois society are so much a part of it that this formulation is justified' (*infra*, p. 82).

3. *Kant's Political Writings*, trans. H. B. Nisbet, ed. H. Reiss, Cambridge, 1970, p. 56.

4. *Werke*, ed. Cassirer, Berlin, 1912, vol. VIII, p. 227.

except freedom, the freedom 'to make *public* use of one's reason in all matters'.[5] The freedom which confronts authority thus has a public character; it is only through this that it enters the concrete dimension of social existence; authority and freedom meet within *bourgeois society* and are posed as problems of bourgeois society. The contradiction is no longer between the 'inner' freedom of the Christian man and divinely ordained authority, but between the 'public' freedom of the citizen and bourgeois society's interest in discipline. Kant's solution remains dualistic; his problematic is in parallel with Luther's: 'the *public* use of man's reason must always be free, and this alone can bring about enlightenment among men; the *private* use of the same may often be very strictly limited, yet without thereby particularly hindering the progress of enlightenment'.[6] That seems to be the exact opposite of Luther's solution, which, while unconditionally preserving the 'inner' freedom of the private person, had also unconditionally subordinated public freedom to the worldly authority. But let us see what Kant means by the 'public' and 'private' use of freedom. 'But by the public use of one's own reason I mean that use which anyone may make of it *as a man of learning* addressing the entire *reading public*. What I term the private use of reason is that which a person may make of it in a particular *civil* post or office with which he is entrusted.'[7] What is 'private' is now the bourgeois 'office', and its bearer has to subordinate his freedom to society's interest in discipline. Freedom in its unrestricted, public nature, on the other hand, is shunted off into the dimension of pure scholarship and the 'world of readers'. Social organization is privatized (the civil 'office' becomes a private possession) and in its privatized form appears as a world of disciplined, controlled freedom, a world of authority. Meanwhile the 'intellectual world' is given the appearance of being actually public and free but is separated from public and free *action*, from real social praxis.

Kant places the problem of authority and freedom on the foundation of the actual social order, as a problem of 'bourgeois

5. *Kant's Political Writings*, p. 55.   6. Loc. cit.   7. Loc. cit.

society'. Even if this concept is by no means historically defined in Kant, but signifies the overall 'idea' of a social order (as a 'legal order'), the actual features of bourgeois society are so much a part of it that the above formulation is justified. We must examine Kant's explication of bourgeois society more closely in order to describe adequately his attitude to the problem of authority. It is to be found in the first part of the *Metaphysics of Morals*, in the *Metaphysical Elements of the Theory of Law*.

Bourgeois society is, for Kant, the society which 'safeguards Mine and Thine by means of public laws'.[8] Only in a bourgeois context can there be an external Mine and Thine, for only in this context do public laws 'accompanied by power' guarantee 'to everyone his own';[9] only in bourgeois society does all 'provisional' acquisition and possession become 'peremptory'.[10] Bourgeois society essentially achieves this legally secure position for the Mine and the Thine in its capacity as 'legal order', indeed, it is regarded as the 'ultimate purpose of all public right' to ensure the peremptory security of the Mine and Thine.

What then is 'right', this highest principle of the bourgeois order? Right is 'the sum total of those conditions under which the will of one person can be united with the will of another in accordance with a universal law of freedom'.[11] All formulations of Kant's concept of right signify a synthesis of opposites: the unity of arbitrary will and right, freedom and compulsion, the individual and the general community. This synthesis must not be thought of as a union which is the sum of individual 'parts'; instead, one should 'see the concept of right as consisting immediately of the possibility of combining universal reciprocal coercion with the freedom of everyone'.[12]

'Only the external aspect of an action'[13] is subject to right in Kant's view. The person as a 'moral' subject, as the locus of transcendental freedom, stands entirely outside the dimension of right. But the meaning of right here is the order of bourgeois

8. *Werke*, vol. VII, p. 44. 9. Op. cit., p. 59.
10. Op. cit., p. 68, and *Kant's Political Writings*, p. 163.
11. *Kant's Political Writings*, p. 133.
12. *Kant's Political Writings*, p. 134. 13. Loc. cit.

society. Transcendental freedom only enters into the legal order in a very indirect way, in so far as the universal law of rights is meant to counteract certain hindrances to the 'manifestations' of transcendental freedom.[14] With this relegation of law to the sphere of 'externality', both law and the society ordered by law are relieved of the responsibility for 'actual' freedom and opened up for the first time to unfreedom. In the synthesis of law we thus have the concerns of the 'externally' acting man before us; what do they look like?

We see a society of individuals, each one of whom appears with the natural claim to the 'free exercise of his will', and confronts everyone else with this claim (since the field of possible claims is limited); a society of individuals, for each one of whom it is a 'postulate of practical reason' to have as his own very external object of his will[15] and who all, with equal rights, confront each other with the natural striving after 'appropriation' and 'acquisition'.[16] Such a society is a society of universal insecurity, general disruption and all-round vulnerability. It can only exist under a similarly universal, general and all-round order of coercion and subordination, the essence of which consists in securing what is insecure, stabilizing what is tottering and preventing 'lesions'. It is highly significant that almost all the basic concepts of Kant's theory of right are defined by negative characteristics like securing, lesion, restriction, prevention and coercion. The subordination of individual freedom to the general authority of coercion is no longer 'irrationally' grounded in the concupiscence of the 'created being' and in the divinely ordained nature of government, but grows immanently out of the requirements of bourgeois society – as the condition of its existence.

But Kant still feels the contradiction between a society of universal coercion and the conception of the 'naturally' free individual. The synthesis of freedom and coercion must not occur in such a way that the original freedom of the individual is sacrificed to social heteronomy. Coercion must not be brought to the

14. Op. cit., p. 133. Cf. Haensel, *Kants Lehre vom Widerstandsrecht*, Berlin, 1926, pp. 10ff.   15. *Werke*, vol. VII, p. 48.   16. Op. cit., p. 70.

individual from without, the limitation of freedom must be a self-limitation, the unfreedom must be voluntary. The possibility of a synthesis is found in the idea of an original 'collective-general' will to which all individuals agree in a resolution of generally binding self-limitation under laws backed by power. That this 'original contract' is only an 'Idea'[17] needs no further discussion, but before we examine its content we must note the significance of its 'ideal' character for the development of the problem under discussion.

Firstly it transforms the historical facticity of bourgeois society into an *a priori* ideal. This transformation, which is demonstrable in Kant's theory of right at the very moment of its occurrence, does not simply mean the justification of a particular social order for all eternity; there is also at work in it that tendency towards the transcendence of the bourgeois authority-system which had already emerged in the Reformers of the sixteenth century. These destructive moments appear in the replacement of a (believed and accepted) fact by a (postulated) 'as if'. For Luther, divinely ordained authority was a given fact; in Kant the statement 'All authority is from God' only means we must conceive of authority '*as if*' it did not come from men, 'but none the less must have come from a supreme and infallible legislator'.[18] Correspondingly, the idea of a 'general will' only requires that every citizen be regarded 'as if he had consented within the general will'.[19] Certainly the 'transcendental As If' signifies a marked shift in the weight of authority towards its free recognition by the autonomous individual, and this means that the structure of authority has become rational – but the guarantees which are set up within the legal order itself against the destruction of the authority relationship are correspondingly stronger.

The 'original contract' is, so to speak, a treaty framework into which the most varied social contents are inserted. But this

17. *Translator's note :* 'Idea' is used here in the Kantian sense of a regulative principle of reason not found in experience but required to give experience an order and unity it would otherwise (according to Kant) lack.

18. *Kant's Political Writings*, p. 143 (Marcuse's emphasis).

19. Op. cit., p. 79.

multiplicity of elements is centred on one point; on the universal, mutual effort to make possible and secure 'peremptory' property, the 'external Mine and Thine', on the 'necessary unification of everyone's private property'.[20] In this way the mere 'fortuitousness' and arbitrariness of 'empirical' property is transformed into the legal validity and regularity of 'intelligible' property in accordance with the postulate of practical reason.[21] We must briefly follow this road through its most important stages, for it is at the same time the route towards the foundation of (social) authority.

Our starting-point is the peculiar (and defining) character of private property as something external, with which 'I am so connected that the use which another would like to make of it without my permission would injure me'.[22] The fact that someone else can use something possessed by me at all presupposes a very definite divorce between the possession and its possessor, presupposes that property does not merely consist in physical possession. The actual 'technical explanation' of the concept of 'private property' must therefore include this feature of 'property with physical possession': 'that which is externally mine is that which, if I am hindered in its use, would injure me, *even if I am not then in possession of it* (if the object is not in my hands)'.[23] What type of property is this property 'even without possession', which is the real subject dealt with by the legal order?

The separation of empirical and intelligible property lies at the basis of one of Kant's most profound insights into the actual structure of bourgeois society: the insight that all empirical property is essentially 'fortuitous' and is based on acquisition by 'unilateral will' ('appropriation') and thus can never present a universally binding legal title; 'for the unilateral will cannot impose on everyone an obligation which is in itself fortuitous. . .'.[24] This empirical property is not therefore sufficient to justify its all-round and lasting security at the centre of the

20. *Werke*, vol. VI, p. 130.   21. Ibid., vol. VII, paras 6, 7 and 11.
22. Ibid., p. 47.   23. Ibid., p. 51.
24. Ibid., pp. 66ff.

bourgeois legal order; instead of this, the possibility of an external Mine and Thine as a 'legal relationship' is 'completely based on the axiom that a purely rational form of property without possession is possible'.[25]

The way in which Kant constructs this axiom and in which he effects the return from empirical property to a 'purely rational form of property' in many ways corresponds to bourgeois sociology's handling of the problem. Kant says: 'In order to be able to extend the concept of property beyond the empirical and to be able to say that every external entity subjected to my will can be counted as mine by right if it is . . . in my power without being in possession of it, all conditions of the attitude which justifies empirical property must be eliminated [ignored] . . .';[26] the 'removal of all empirical conditions in space and time', abstraction from the 'sensuous conditions of property'[27] leads to the concept of 'intellectual appropriation'. By this route Kant arrives at the idea of an original joint ownership of the land and on the basis of this collectivity a collective general will can be established which legally empowers every individual to have private property. 'The owner bases himself on the innate *communal ownership* of the land and a general will which corresponds *a priori* to this and allows *private ownership* on the land. . . .'[28] Thus in a highly paradoxical manner communal property becomes the 'legal basis' for private property; total ownership 'is the only condition under which it is possible for me to exclude every other owner from the private use of the object in question. . . .'[29] No one can oblige anyone else through unilateral will to refrain from the use of an object: the private appropriation of what is universal is only possible as a legal state of affairs through the 'united will of all in total ownership'. And this 'united will' is then also the foundation of that general community which puts every individual under a universal coercive order backed by force and which takes over the defence, regulation and 'peremptory' securing of the society based on private property.

25. Ibid., p. 57.   26. Ibid., p. 54.   27. Ibid., pp. 67 and 72.
28. Ibid., p. 52.   29. Ibid., p. 64.

Thus in the origins of bourgeois society the private and general interest, will and coercion, freedom and subordination, are meant to be united. The bourgeois individual's lack of freedom under the legal authority of the rulers of his society is meant to be reconciled with the basic conception of the essentially free person by being thought of as the mutual self-limitation of all individuals which is of equally primitive origin. The formal purpose of this self-limitation is the establishment of a general community which, in uniting all individuals, becomes the real subject of social existence.

'The general community' is society viewed as the totality of associated individuals. This in turn has two connotations:

1. A *total communality* of the kind that reconciles the interests of every individual with the interests of the other individuals – so that there is really a general interest which supersedes private interests.

2. A *universal validity* of such a kind that the general interest represents a *norm equally binding* on all individuals (a law). In so far as the interests of the individuals do not prevail 'on their own', and do not become reconciled with each other 'on their own' (in a natural manner), but rather require social planning, the general community confronts the individuals as a priority and as a *demand*: in virtue of its general 'validity' it must demand recognition and achieve and safeguard this by coercion if necessary.

But now everything depends on whether the general community as the particular form of social organization does in fact represent a supersession of private interests by the general interest, and whether the people's interests are really guarded and administered in it in the best possible way. When Kant deals with social problems in the context of the 'general community', this already signifies a decisive step in the history of social theory: it is no longer God but man himself who gives man freedom and unfreedom. The unchaining of the conscious bourgeois individual is completed in theory: this individual is so free that he alone can abrogate his freedom. And he can only be free if at the same time freedom is taken away from all others: through all-round, mutual subordination to the authority of the law. The bearer of authority (in the sense of being the source of authority) is not God, or a person or a multiplicity of persons, but the general community of all (free) persons in which every individual is both the person delegated and the person delegating.

But not every general community, i.e. every actually constituted society, is truly universal. German Idealism uses bourgeois society as a model for its exposition of the concept of universality: in this sense, its theory signifies a new justification of social unfreedom. The characteristics of real universality are not fulfilled in this society. The interests of the ruling strata stand in contradiction with the interests of the vast majority of the other groups. The universally obligatory authority of the law is thus finally based not on a 'genuine' universality (in which the interests of all the individuals are common to all) but on an appearance of universality; there is an apparent universality because the particular interests of certain strata assume the character of general interests by making themselves apparently independent within the state apparatus. The true constituents of this universality are property relationships as they existed at the 'beginning' of bourgeois society and these can only be peremptorily guaranteed through the creation of a universally binding organization of social coercion.

This universality retains its 'private' character; in it the opposing interests of individuals are not transcended by the interests of the community but cancelled out by the executive authority of the law. The 'fortuitousness' of property is not eliminated by the 'elimination' of the empirical conditions under which it was appropriated: right rather perpetuates this fortuitous character while driving it out of human consciousness. The universality which comes from the combination of private possessions can only produce a universal order of injustice. Kant knew that he had constituted his theory of right for a society whose very foundations had this inbuilt injustice. He knew that 'given man's present condition ... the good fortune of states grows commensurably with the misery of men',[30] and that it must

30. Ibid., vol. VIII, pp. 465ff.

be a 'principle of the art of education' that 'children should be educated not towards the present, but towards the future, possibly better, conditions of the human race'.[31] He has said that in this order justice itself must become injustice and that 'the legislation itself (hence also the civil constitution), so long as it remains barbarous and undeveloped, is to blame for the fact that the motives of honour obeyed by the people are subjectively incompatible with those measures which are objectively suited to their realization, so that public justice as dispensed by the state is *injustice* in the eyes of the people'.[32]

None the less Kant stuck to the view that the universality of the 'united will' was the basis of society and the foundation of authority. He drew all the resultant consequences from the unconditional recognition of the government ruling at any particular time to the exclusion of economically dependent individuals from civil rights.[33] Like Luther he maintained that right was immanent in the civil order and described rebellion against this order as the 'overthrow of all right',[34] and as 'the road to an abyss which irrevocably swallows everything',[35] the road to the destruction of social existence altogether. 'There can thus be no legitimate resistance of the people to the legislative head of state; for a state of right is only possible through submission to his universal legislative will....'[36] His justification is in the first place purely formal: since every existing system of domination rests only on the basis of the presupposed general will in its favour, the destruction of the system of domination would mean the 'self-destruction' of the general will. The legal justification is of the same formal kind: in a conflict between people and sovereign there can be no tribunal which makes decisions having the force of law apart from the sovereign himself, because any such tribunal would contravene the 'original contract'; the sovereign is and remains, says Kant in a characteristic phrase, in sole

31. Ibid., vol. VIII, pp. 462ff. 32. *Kant's Political Writings*, p. 159.
33. Op. cit., pp. 139ff.; p. 78. 34. Op. cit., p. 162.
35. Op. cit., p. 146 (Kant's footnote to paragraph 49).
36. Op. cit., pp. 144ff.; other important passages are in op. cit., p. 143, pp. 81-2, pp. 126-7, and in *Werke*, vol. VII, pp. 179ff.

'possession of the ultimate enforcement of the public law'.[37] This is the consequence of the immanence of the law in the ruling system of authority already observed in Luther: the sovereign is his own judge and only the judge himself can be the plaintiff: 'Any alteration to a defective political constitution, which may certainly be necessary at times, can thus be carried out only by the sovereign himself through *reform*, but not by the people, and, consequently, not by *revolution*. . . .'[38]

It has been pointed out in connection with Kant's strict rejection of the right of resistance that although he does not acknowledge a (positive) 'right' of resistance as a component of any conceivable legal order, the idea of possible resistance or even of the overthrow by force of a 'defective' social order, is fully in line with his practical philosophy. The main support for this interpretation (which can be reconciled with the wording of the quoted passages of his theory of right) is Kant's apotheosis of the French Revolution in the *Contest of the Faculties*,[39] and the unconditional demand for the recognition of every new order arising from a revolution.[40] Such an interpretation strikes us as correct, as long as it does not attempt to resolve the contradiction present in Kant's position in favour of one side or the other. The transcendental freedom of man, the unconditional autonomy of the rational person, remains the highest principle in all dimensions of Kant's philosophy; here there is no haggling and calculating and no compromise. This freedom does not become a practical social force, and freedom to think does not include the 'freedom to *act*';[41] this is a feature of precisely that social order in the context of which Kant brought his philosophy to concreteness.

The internal antinomy between freedom and coercion is not resolved in the 'external' sphere of social action. Here all freedom remains a state of merely free existence under 'coercive laws', and each individual has an absolutely *equal* inborn right 'to coerce others to use their freedom in a way which harmonizes with his

37. *Kant's Political Writings*, p. 82.
39. Op. cit., pp. 182–5.
41. Op. cit., p. 59.

38. Op. cit., p. 146.
40. Op. cit., p. 147.

freedom'.[42] But mere self-subordination to general coercion does not yet provide the foundation for a generality in which the freedom of individuals is superseded. On the road from empirical to intelligible property, from the existent social universality to the Idea of an original universality, the solution of the antinomy is transferred to the transcendental dimension of Kant's philosophy. Here too the problem appears under the heading of a universality in which the freedom of the individual is realized within a general system of legislation.

In the 'external' sphere the relationship between freedom and coercion was defined in such a way that coercion was made the basis of freedom, and freedom the basis of coercion. This notion is most pregnantly expressed in the formula which Kant uses in his discussion of a 'purely republican' constitution: it is the only state form 'which makes *freedom* into the principle, indeed the *condition*, of all coercion'.[43] Just as 'legitimate' coercion is only possible on the basis of freedom, so 'legitimate' freedom itself demands coercion in order to survive. This has its rationale within the 'external' sphere: 'bourgeois' freedom (this is what is at stake here), is only possible though all-round coercion. But the result is not a supersession but a reinforcement of actual unfreedom: how then can this be reconciled with transcendental freedom?

The concept of transcendental freedom (the following discussion will be limited to this, unless otherwise indicated) appears in Kant as a concept of causality. This concept stands in opposition to that of causality in nature: it refers to causality resulting from free actions as opposed to causality resulting from necessity and its causal factors, which are of 'external' origin (i.e. causality in the sequence of temporal phenomena). People have seen in this definition of freedom as a type of causality an early derivation of the problem of freedom – a dubious transference of the categories of natural science into the dimension of human existence, and a failure to understand the 'existential' character of human freedom. But we believe that what shows the superiority of Kant's ethics over all later existential ontology is precisely this understanding

42. Op. cit., p. 76.   43. Op. cit., p. 163.

of freedom as, from the start, a particular type of actual effectiveness in the world; freedom is not relegated to a static mode of
existence. And since the definition of causation resulting from
freedom has to meet from the outset the demand for 'universal
validity' and since the individual is placed in a universal, a general
rational realm of free persons which exists 'before' and 'over' all
natural aspects of the community, all later misinterpretations of
the organicist theory of society are refuted from the start. However, freedom is now set up as unconditional autonomy and pure
self-determination of the personal will, and the required universal
validity is posited as *a priori* and formal: here we see the impact
of the inner limits of Kant's theory of freedom (and these limits
are by no means overcome by proposing a 'material ethic of value'
as against 'formal' ethics).

Freedom for Kant is a transcendental 'actuality', a 'fact'; it is
something which man always already has if he wants to become
free. As in Luther, freedom always 'precedes' any free act, as its
eternal *a priori*; it is never the result of a liberation and it does not
first require liberation. Admittedly freedom 'exists' for Kant only
in activity in accordance with the moral law, but this activity is, in
principle, free to everyone everywhere. By the ultimate reference
of freedom to the moral law as its only 'reality', freedom becomes
compatible with every type of unfreedom; owing to its transcendental nature it cannot be affected by any kind of restriction
imposed on actual freedom. Admittedly freedom is also a liberation – man making himself free from all 'empirical' determinants
of the will, the liberation of the person from the domination of
sensuality which enters into the constitution of the human animal
as a 'created being' – but this liberation leaves all types of actual
servitude untouched.

The self-imposed and self-observed moral law of the free
person possesses 'universal validity' in itself as the reason of
knowing of its truth, but this means that it contains reference to a
'world' of universality consisting of the mutual coexistence of
individuals. Nevertheless, this universality is formal and aprioris-
tic; it may not carry over anything of the material quality of this

mutual coexistence into the law of action. Yet another 'form' is concealed in the bare 'form' of the moral law; namely the bare form of the coexistence of individuals, the form of a 'society as such'. This means that in all his actual decisions about action the individual only has the form of social existence in view: he must disregard or, so to speak, leap over the social materiality before him. Precisely to the extent that the individual acts under the law of freedom can no element of this materiality be permitted to become a determinant of his will. The fact that it is entirely excluded from the determinants of free praxis means that the individual comes up against it as a brute fact. Transcendental freedom is by its nature accompanied by social unfreedom.

The criterion for decisions concerning action under the moral law is, as already in the sphere of the theory of right, the internal coherence of maxims as a universal law: a bad maxim, if it were made into a 'universal system of legislation', would abolish the order of human coexistence; it would signify the self-destruction of social existence. It has already been shown elsewhere that this criterion cannot operate in the intended sense in a single one of the applications which Kant himself adduces.[44] It would not be the form of a social order as such which would be destroyed by 'false' maxims but always only a particular social order (Kant's ethics are by no means as formal as is claimed by the material ethics of value). Between the formal universality of the moral law and its possible universal material validity, there yawns a contradiction which cannot be overcome within the Kantian ethic. The existing order, in which the moral law is meant to become a practical reality, is not a field of real universal validity. And the alteration of this order cannot in principle serve as a maxim of free praxis, for it would in actual fact, judged according to Kant's criterion, transcend social existence as such (a universal law for the alteration of the existing order would be an absurdity).

The reversion from personal and institutional authority to the authority of the law corresponds to the justificatory reversion from the subject-matter of praxis to the form of the 'law'. This

44. *Zeitschrift für Sozialforschung*, II (1933), pp. 169ff.

'formalization' is something quite different from Luther's 'formal' recognition of the existing wordly authorities, without reference to their individual and social basis. For Kant, every personal and institutional authority has to justify itself in face of the idea of a universal law, which the united individuals have given themselves and which they themselves observe. In the 'external' sphere of social existence this law – as we have seen in the theory of right – justifies not only the authority of the actual system of 'governments' but also authority in general as a social necessity; universal voluntary self-limitation of individual freedom in a general system of the subordination of some and the domination of others is necessary for the peremptory securing of bourgeois society, which is built up on relations of private property. This is the highest rationalization of social authority within bourgeois philosophy.

But just as, with the application of the law, rationalization is brought to a standstill in face of the internal contradictions of bourgeois society, in face of its immanent 'injustice', so it is with the origin of legislation itself: 'the possibility of an intelligible property, and thus also of the external Mine and Thine, is not self-evident, but must be deduced from the postulate of practical reason.'[45] The law remains an authority which right back to its origins cannot be rationally justified without going beyond the limits of precisely that society for whose existence it is necessary.

45. Op. cit., vol. VII, pp. 57f.

Kant had introduced the antagonism between freedom and coercion into the idea of freedom itself: there is only freedom under the (coercive) law. The supersession of this antagonism was sought in the unification of the individual and the general community. In the sphere of social action this appeared as the voluntary all-round self-limitation of the united individuals through which social existence as a world of free individuals or as 'bourgeois society' became possible for the first time.

The 'universality' which lies at the basis of bourgeois society is by no means able to fulfil its function of replacing individual freedom with a general freedom; this fact is the starting point for Hegel's critique of Kant's theory of law: 'Once the principle is adopted that what is fundamental, substantive, and primary is the will of a single person ... a particular individual ... in his own private self-will ..., the rational can of course only come on the scene as a restriction on the type of freedom which this principle involves ... and only as an external abstract universal.'[1] The problem of freedom in Hegel remains subject to the idea of universality:[2] individual freedom can only become real in a 'general community'. The task is to define this universality conceptually and to indicate its social reality.

The description of bourgeois society in Hegel's philosophy of law is completely based on the recognition that the universality which has come into being in this society does not represent a 'true' universality and thus not a real form of freedom (realized through its supersession). Moreover it *cannot* represent this, so

1. *Philosophy of Right*, trans. T. M. Knox, Oxford, 1952, para. 29.
2. *Translator's note :* German: *Allgemeinheit*. 'Universality' is the usual rendering of this philosophical concept, but where the German refers to the concrete political form of the concept, i.e. the mass of individuals bound together in a community or state, the phrase 'general community' has been used.

that the realization of true freedom necessarily leads beyond bourgeois society as such.

The double 'principle' of bourgeois society is 'the concrete person, who is himself the object of his particular aims. . . . But the particular person is essentially so related to other particular persons that each establishes himself and finds satisfaction by means of the others, and at the same time purely and simply by means of the form of universality. . . .'³ The particular person himself in this society is only a 'mixture of natural necessity and arbitrariness'; the clashing of 'selfish ends' produces a 'system of universal dependence', which may be able to 'safeguard' the subsistence, well-being and rights of the individual but as a whole continues to be governed by 'external accident and caprice'.⁴ The general community is, to begin with, nothing more than the mutual dependence of 'selfish' individuals, a world of private satisfaction of needs. 'The individuals as citizens of this state are *private persons* whose end is their own interest. This end is *mediated* through the universal which thus *appears* as a *means* to its realization.'⁵ The principle of this 'system of needs' only contains this universality of freedom 'abstractly, that is, as the *right of property* which, however, is no longer merely implicit but has attained its recognized actuality as the *protection of property* through the administration of justice'.⁶ The highest stage of the unity of subjective particularity and universality which can be reached by such an order of universal fortuitousness and arbitrariness is thus a primary organization of coercion and interests: 'the actualization of this unity through its extension to the whole ambit of particularity, is (i) the specific function of the Police, through the unification which it effects is only relative; (ii) it is the Corporation which actualizes the unity completely, though only in a whole which, while concrete, is restricted.'⁷

Hegel sees civil society basically from the same viewpoint as Kant: as a universal coercive order for the safeguarding (of the

---

3. *Philosophy of Right*, para. 182.
4. Ibid., para. 185.　　　　5. Ibid., para. 187.
6. Ibid., para. 208.　　　　7. Ibid., para. 229.

property of) free private property owners – an order whose authority may be 'universal' (its claims being recognized by all the individuals organized within it because of their own interests) and legitimate, but which stands and falls with its own basis and presupposition: namely a social order for the peremptory safeguarding of private property. Kant saw this presupposition as necessary for any idea of a 'legitimate' social order; Hegel does not contradict him in this. But in contrast to Kant his picture of bourgeois society is coloured by its negativity. When bourgeois society is in a state of 'unimpeded activity', 'the amassing of wealth . . . is intensified on the one hand, while the subdivision and restriction of particular jobs, and thus the dependence and distress of the class tied to work of that sort, increases on the other hand'.[8] For the first time the revolutionary character of the dialectic breaks through into the dimension of civil society: the image of this society, which was still essentially static in Kant, begins to move. Despite all the 'excess of wealth', civil society is not rich enough to 'check excessive poverty and the creation of a penurious rabble';[9] through this dialectic it is 'driven beyond its own limits'.[10] Where to? The dialectic evades the real answer to this question by withdrawing into the house of the philosophical system; the relevant passages in the *Philosophy of Right* merely point to the world economic market and colonization as a way out. The systematic continuation of the dialectic is something different: it leads to the supersession of civil society by the 'state'. The idea of civil society itself constituting itself as a state is rejected; society and state are separated according to their 'principle'. This is a decisive step for the development of the problem of authority: civil society, now seen almost in its full problematic, can no longer in itself provide the basis for the social system of authority; it ceases to be the real basis of freedom and thus also of the universal community. The state confronts it as an independent totality and is thus liberated from its negativity and becomes the unconditional bearer of all social authority. The

8. Ibid., para. 243. 9. Ibid., para. 246.
10. *Early Theological Writings*, trans. by T. M. Knox, Chicago, 1948, p. 221.

thorough-going rationalization of the authoritative order is abandoned; the philosophy of absolute reason sets up a completely irrational authority on the foundations of the state. That is a rough outline of what we must now examine in detail as the form taken by the problem of authority in Hegel's philosophy of the state.

Hegel, like Kant, sees state and society in the context, first and foremost, of the idea of property. As early as 1798–99 he says with regard to Jesus's call to 'cast aside care for one's life and despise riches': 'It is a litany pardonable only in sermons and rhymes, for such a command is without truth for us. The fate of property has become too powerful for us to tolerate reflections on it, to find its abolition thinkable.' And his work on the German Constitution makes this a straight question of definition: 'A multitude of human beings can only call itself a state if it is united for the communal defence of the entirety of its property.'[11] But it is precisely here, in the inter-relationship of state and property, that the change begins: the task of legally and politically safeguarding property is taken away from the state as such and transferred to 'civil society' itself; and it is this that brings about the elevation of the state to an independent position with respect to society. 'If the state is confused with civil society and its specific end is laid down as the security and protection of property and personal freedom, then the interest of the individuals as such becomes the ultimate end of their association. . . . But the state's relation to the individual is quite different from this.'[12]

Before we look at Hegel's positive definition of this relationship, we must briefly examine the distinction which appears here. Hegel combines the basic separation of state and society with the critique of Rousseau's and Kant's 'contract theory'; since this theory understands the general will merely as the communal aspect of the individual will, the universality of the state is, as it were, privatized; it is reduced to a combination of private persons, 'based on their arbitrary wills, their opinions, and their capri-

11. *Hegel's Political Writings*, trans. T. M. Knox, Oxford, 1964, p. 154.
12. *Philosophy of Right*, para. 258; cf. para. 100.

ciously given express consent; and there follow ... logical inferences which destroy the absolutely divine principle of the state, together with its majesty and absolute authority'.[13] The contract theory transfers 'the characteristics of private property into a sphere of a quite different and higher nature',[14] and such reflections must destroy the absolute authority of the 'divine' state. This is a clear indication of a new tendency to revile the genetic view that the state originates from the (material) interests and needs of individuals as being destructive of authority, and to elevate the objectivity of the state, which 'exists in and for itself' above all empirical conditions. The reasons for the authority of a 'real state', 'in so far as it has anything to do with reasons' can only be taken from 'the forms of the law authoritative within it'.[15] The fear of an historical return to the legal basis of the existing order of the state comes through clearly enough: 'In any case, however, it is absolutely essential that the constitution *should not be regarded as something made*, even though it has come into being in time. It must be treated rather as something simply existent in and by itself, as divine therefore and constant, and so as exalted above the sphere of things that are made.'[16] Faced with 'something existing in and for itself' the question of the legal basis of its authority is already meaningless. The whole series of problems which in Kant's philosophy of the state come under the title of the right of resistance no longer worries Hegel at all. 'Every nation ... has the constitution appropriate to it and suitable for it.'[17] As far as the concept of 'the people' corresponds to any reality at all, other than the general community of the 'ruled' brought together in the state, it can only be used to describe that section of the citizens 'which does *not* know what it wants'.[18]

What happens in this theory of the state at first appears to be a total reification of the social and political orders. The state, which becomes the sole bearer of the authority of this order, is deprived of any historical genesis as a preceding 'totality' and a 'reality

---

13. Ibid., para. 258.
15. Ibid., para. 258.
17. Ibid., para. 274.

14. Ibid., para. 75.
16. Ibid., para. 273.
18. Ibid., para. 301.

existing in and for itself', and is presented as a sphere independent of individual and society. The systematic dialectic, which merges civil society into the state, silences the historical dialectic. The 'sovereignty' of the state, freed from any personal or social basis, appears as a 'metaphysical' quality, peculiar to the state as such: it 'has its ultimate roots ultimately in the unity of the state as its simple self'.[19] This concept of the 'sovereignty of the state purely as such, without express relation to its human bearers' subsequently became the decisive theoretical weapon.[20] The elevation of the sphere of the state above bourgeois society makes it possible to subordinate all social authority to the authority of the state as such. The more obviously bourgeois society loses the appearance of 'real' universality and allows the antagonisms of class society to break through, the less this society can be claimed to be the true supersession of the freedom of the individual. Kant's practical philosophy is refuted by the history of society. It is not the least aim of Hegel's brilliant critique of the Kantian ethic[21] to show the practical impossibility of the social universality proclaimed by Kant and to discover a different, no longer social, universality as the bearer of authority and as the locus of the supersession of individual freedom. Hegel can go so far as to contend that the sphere of private property, 'the interest in proving that property must be', is the presupposition of the Kantian ethic and to use this interest as an example of the emptiness of the laws made by practical reason. For he already has the separation of civil society and the state in view, and this enables him to develop a sphere of universality which appears to be basically separated from this sphere of property. We must now show how Hegel gives a positive definition of this universality, which in his work becomes the real bearer of all authority. For this purpose it is necessary to go into Hegel's transformation of the bourgeois concept of freedom, which gives this concept its decisive form for the subsequent period.

19. Ibid., para. 278.
20. Rosenzweig, *Hegel und der Staat*, Munich, 1920, vol. II, p. 143.
21. *Schriften zur Politik und Rechtsphilosophie*, ed. Lasson, Leipzig, 1911, pp. 355f.; *Philosophy of Right*, paras 29, 135.

The undifferentiated indeterminacy of the will in the freedom of choice, the possibility of abstracting the will from every determinate state of mind, and even Kant's positive concept of the autonomy of the will, as far as this refers only to a 'formal self-activity', belong for Hegel to the merely negative or merely abstract concepts of freedom.[22] This means that the universality which comes into being through the limitation and supersession of such freedom cannot be the true universality. Hegel demands that the concept of freedom be taken out of the dimension of mere feelings, inclination and arbitrary will, and also out of the mere realm of the Ought; there is freedom only in existence, reality, known and conscious reality, 'spirit' (*Geist*). But since freedom is at the same time the substantial definition of the human will, unified with 'intelligence' in the unity of the 'theoretical and practical spirit', it follows that man can only give the existence of freedom to himself: there is freedom only in the free act of man.

The definition of freedom as reality seems to indicate that Hegel is giving the concept of freedom a strongly concrete content. But, in that freedom is still explicated as 'absolute' freedom, a change occurs: in the sphere of actual freedom everything alien, contradictory, external, fortuitous must be superseded; it is without any contradiction (for any contradiction would make it dependent) and thus also no longer has necessity pitted against it: 'freedom, shaped into the actuality of a world, receives the form of necessity'.[23] This essential unity of freedom finds its subjective fulfilment in the fact that the will is constantly 'at home with itself' and its objective fulfilment in the supersession of the tension or, as the case may be, the opposition between the 'concept and the object'. When man, with every single actual determination of his existence, freely determines himself towards this determination and freely acknowledges the necessity which he finds before him, the 'fortuitousness and limitation of the

22. Ibid., paras 5, 15, 29.
23. *Enzyklopädie*, vol. III; *Philosophie des Geistes* (trans. W. V. Wallace as *Hegel's Philosophy of Mind*, Oxford, 1894), para. 484; cf. *Philosophie der Weltgeschichte*, ed. Lasson, vol. I, Leipzig, 1917, p. 94.

previous practical content' are superseded.[24] The will which revolts against reality and is tensed against existence is not yet absolutely free: it is still confronted with something not yet overcome, something external; it is not yet 'with itself'. The truly free will is related to 'nothing except itself' and is thus released from any 'tie of dependence on anything else'.[25]

It is precisely this absorption of all particular individuality and restriction by the will in its state of self-identity which constitutes that 'universality' into which Hegel's theory of freedom debouches.[26] As in Kant, the concept of freedom in Hegel is linked from the outset with the concept of universality, and in the system's final form the concepts of freedom and universality mean almost the same thing.[27] We should need to return to the basis of Hegel's philosophy to unfold the concept of universality; here we must be content to point out the result. The decisive point is that 'universality' is neither a mere determination of the individual will, nor the universal content of the various combined individual wills. The concept aims rather at an objective spiritual reality, as corresponds with the situation of the problem of freedom within the philosophy of the objective spirit. 'The universal must not be simply what is thought by the individuals, it must be something existent; as such it is present in the state, it is that which is valid.'[28]

The being-with-itself of the free will and the disappearance of the contradiction between freedom and necessity is realized in a real universality in which the tension between concept and object is already discarded as the form of the objective spirit and the existent is already 'rational': in the world of 'morality', or, to be more precise, the world of the state. 'The state is the immediate and more closely defined subject of world history as a whole, and in the state freedom obtains its objectivity and lives in the enjoyment of this objectivity.'[29] The state is 'the actuality of concrete

24. *Enzyklopädie*, vol. III, para. 481.   25. *Philosophy of Right*, para. 23.
26. Op. cit., para. 24.
27. Op. cit., para. 38; *Enzyklopädie*, vol. III, para. 485.
28. *Philosophie der Weltgeschichte*, vol. I, p. 92.   29. Ibid., para. 1, p. 90.

freedom' and the idea of freedom is 'genuinely actual only as the state'.[30]

The state for its part is doubly built into Hegel's philosophy as a particular form within world-historical development and within the development of the system. As the sole locus of the 'rational existence' of man and as the 'realization of freedom', it entered reality at a relatively late stage of historical development: as the Christian-Germanic state of the West. The mode of universality realized in it is historical in its origin: human freedom has a history. But this history is complete after Christianity has brought the idea of the freedom of man as such into the world, the idea 'that man in himself is destined for the highest freedom'.[31] The state, as it is now found to exist by the individual, is the actuality of the rational, and the individual only has to acknowledge it as 'that which is valid'. The substantiality of our being is realized in the state; 'the rational has necessary existence, as being the substance of things, and we are free in recognizing it as the law'.[32]

The authority of the state is thus founded at a level quite beyond the reach of the power of the individual; it is based on the development of a 'world spirit' which has progressed on its road through the centuries up to the truth represented by the state. In the face of this, the question of the actual moral basis of authority and the correspondence of the actually given socio-political formation with the needs of man becomes meaningless: 'concept and object' are already united in the state. Freedom can no longer become objective, since it always already is: the idea of freedom is 'the actuality of men, not something which they have, as men, but which they are'.[33] If freedom has thus become actual in the universality of the state, then the freedom of the individual can only consist in the absorption of his 'arbitrary' independence by this universality; the independence of individuals is present 'only in the state'. 'The individual obeys the laws and knows that he has

30. *Philosophy of Right*, paras 260, 57.
31. *Enzyklopädie*, para. 482.
32. *Philosophie der Weltgeschichte*, vol. I, p. 94.
33. *Enzyklopädie*, vol. III, para. 482.

his freedom in this obedience.'[34] Perhaps for the first time in bourgeois philosophy *history* becomes the first and final authority, but through the closed nature of the system a particular form of historical development is posited as absolute: the 'understanding of that which is' acquires the quietistic tone of a justificatory recognition of the existing situation.

The most extreme point of the subordination of the individual to the authority of the state, which, in its universality, continues according to Hegel to 'correspond' to him, has its counterpart on the other side, at the very summit of the state, in a completely 'groundless' and 'unmediated' authority: the authority of the monarch. The ultimate self in which the will of the state is concentrated, 'raised above all particularity and conditions', no longer bases its authority on history but on – 'Nature'. This concept contains the 'definition of naturalness': the dignity of the monarch is determined 'in an immediate natural fashion, through his birth in the course of nature'.[35] Hegel simply piles up characteristics which emphasize the irrational nature of the hereditary monarchy: 'the will's ungrounded self', the 'ungrounded immediacy' and the ultimate 'being in itself' which does not risk being drawn down 'into the sphere of capricious argument' and which, precisely because of its irrational naturalness, is excluded from the conflict of factions around the throne and 'from the enfeeblement and overthrow of the power of the state'.[36] The recourse to ungrounded naturalness as the last protection of authority is not the only place in which irrationality breaks into this system of reason. Before we return to this point, we must examine another tendency of Hegel's philosophy of the state which is important for the problem of authority.

The (subjective) basis of Hegel's philosophy of the state and society is – as already in Kant – the human will: bourgeois society is dealt with as the sphere of existence of the free will, and the state as its completed actuality. The construction of the state out of the will of individuals ended with the free subordination of

---

34. *Philosophie der Weltgeschichte*, vol. I, p. 99.
35. *Philosophy of Right*, para. 280.   36. Ibid., para. 281.

the individual will to the general will of the state. It demands in its turn what might be called a subjective preparation: the building up of a state-upholding sentiment in the psyche of the individual; the authority of the state must be rooted in the basic psychological attitude of the citizen. We shall follow this process in its more important stages, for it can almost be taken as a sketch of the development of the authoritarian consciousness.

The 'institutions' of the state essentially have the effect of producing and continually keeping alive the 'political attitude' which forms the subjective foundation of the state. 'The political sentiment, patriotism pure and simple . . . and a volition which has become habitual, are purely a product of the institutions subsisting in the state. . . .'[37] The institutions of the state, however, which the individual always finds before him in their finished form, are not enough to make this state-upholding volition habitual. The preparation goes further back into the history of the individual: through the state of the 'corporations' to the 'family'. 'As the family was the first, so the corporation is the second ethical root of the state, the one planted in civil society.'[38] In particular, notions of civil qualifications, orderliness and efficiency, and 'rank and dignity' are ways in which the individual is tied to the general community. 'Unless he is a member of an authorized corporation . . . an individual is without rank or dignity.'[39] His civil 'recognition' within the general community presupposes that he himself recognizes the universality of that community's institutions. The significance of that other, prior 'root' of the state – the *family* – is even more basic. We must, however, guard against the misunderstanding to the effect that Hegel assumed a genetic development of the state out of the family (like some sociological theories). Rather the family is for him the 'ethical' root of the state: it brings out characteristics through which the individual can become a part of the state which represents 'objective' morality; it is the first, still direct and natural form of the objective universality which supersedes

37. Ibid., para. 268.   38. Ibid., para. 255.
39. Ibid., para. 253.

'subjective particularity'; it is the 'ethical spirit' in its immediate and natural form.

The features which qualify the family for such a function are: the direct unification of individuals into a general community without the person as such being negated; the real character of this general community of which the individuals are constantly aware in their everyday existence; and the actual communal nature of needs and interests which, since they concern an actual universality, are raised from the sphere of mere selfishness and 'moralized'. But all these features of the family are only realized in that centre around which all features of the family are grouped in Hegel: in the specific relationship between *family* and *property*. The family not only has its 'external reality' in property, but also the existence of its 'substantial personality'. Only in and through the family is property transformed from the 'arbitrary expression of the particular need' to a 'permanent and secure asset', and the 'selfishness of greed' is transformed into 'something ethical, into labour and care for a common possession'.[40]

From this we can see the full significance of the family on the road from individual to state, from egoism to a state-orientation. The individual as an existing person for Hegel is basically – a private owner. Only in property does the person 'become merged with himself',[41] only in property does he possess the 'external sphere of his freedom'. So essential is the inter-relationship of personal freedom and property that property is not only a means for satisfying needs but 'from the standpoint of freedom, property is the first embodiment of freedom and so is in itself a substantive end'.[42] But as long as the individual remains tied to the 'arbitrariness' of private property, it is not possible to realize that actual universality which the socio-political order must possess for its authority. Since the idea of the non-existence of private property (after property has once been proclaimed as 'destiny') cannot even be discussed, a relationship between the individual and the general community must be established in and through property

40. Ibid., para. 170.
41. *Enzyklopädie*, vol. III, para. 490.
42. *Philosophy of Right*, para. 45.

itself: property must in a certain manner be stripped of its merely 'private' and egoistic character without thereby losing the character of property. It is essentially the family which accomplishes this, or more exactly, the right of inheritance of the family. Since the family as a whole, and not the individual, becomes the actual subject of property, coming into inheritance only means entering into the ownership of 'assets which are themselves communal'.[43] The universality of property is safeguarded particularly from the arbitrary will of the person himself, through a characteristic limitation on the arbitrariness of the freedom of bequest. Since property is anchored in the family and guaranteed in the right of inheritance through successive generations, the individual receives his property, as it were, from the general community itself, by force of an eternal natural order, in trust for, and in the service of, the general community. It is the specific function of the family in moralizing and eternalizing property which justifies the elevation of the state above the sphere of property, as revealed in the separation between the state and civil society. Society and state are relieved of the task of the primary 'peremptory' safeguarding of property, since this has already been taken over by the family.[44] In the subsequent period the family, with these functions, enters bourgeois sociology as the basis of the state and society.

In the return from the 'finished' socio-political order, the family is not the final stage on which this order is constructed and the individual integrated into the general community. The further stage of this return leads back to earlier levels of Hegel's philosophy which in the completed system have lost some of their original importance. At these earlier levels, the historical-social world is not yet seen in the later quietistic-justificatory manner: the dialectic has not yet been forced from its ground through

43. Ibid., para. 178.
44. Cf. Rosenzweig's accurate formulation (op. cit., vol. II, p. 118): 'The family's right of inheritance, based on the communal nature of the family income, serves to uphold the necessary connection between person and property without necessitating the direct intervention of the state and society. It is the first and decisive line of demarcation by which Hegel's thesis of property as a necessity for every individual differentiates itself in advance from communism.'

being enclosed into a system, and it thus reveals its full force. We shall pass over the significance of the family in the *Phenomenology of Mind*, and pursue the question of the building up and the anchoring of the authoritative socio-political order back to the genesis of the *Phenomenology*. Here we find the family in close proximity to the relationship of *domination* (*Herrschaft*) and *servitude* (*Knechtschaft*) in which Hegel discerns mutual 'recognition' as the basis of social existence. In his *Jenenser Realphilosophie* of 1805–6, the establishment of the family immediately follows the struggle for property ending in the recognition of property as a general right: and in the *System of Morality* of 1802 the family is the 'external, openly manifested' element of the relationship between domination and servitude in its 'indifference'.[45]

Within the world of the 'spirit', which is the historical-social world, human existence is firstly 'self-consciousness'. But self-consciousness is 'in and for itself' only because it is in and for itself through another, which means 'only in being something recognized'.[46] If recognition is thus placed at the beginning of social order, this concept does not refer merely to the voluntary subordination, somehow deriving from insight, of one person to another which occurs over and above direct force (we shall show how this happens below), but to the justification of such recognition in the material sphere of society: it occurs in Hegel, after a 'life and death struggle', in the realm of appropriation and property, work and service, fear and discipline. The way in which the domination of the master is constituted (and here we put together the expositions of the *Phenomenology* and the stages of the system preceding it) is 'greed' for the 'enjoyment' of things, 'appropriation' as the 'sensuous acquisition of property', through which the other is 'excluded' from ownership, and the binding of the subordinated person through the 'work' which is forced upon

45. For the interpretation of the dialectic of domination and servitude and its systematic place in Hegel's philosophy, cf. H. Marcuse, *Hegels Ontologie und die Grundlegung einer Theorie der Geschichtlichkeit*, Frankfurt, 1932, pp. 291ff. We must limit ourselves here to an outline of this work's conclusions.

46. *The Phenomenology of Mind*, trans. J. B. Baillie, London, 1966, p. 255.

him, in which the servant 'works on' and 'forms' things for the enjoyment of the master. The servitude of the servant is constituted by his material powerlessness, his 'absolute fear' of the master, his constant 'discipline' of service and above all his being chained to his work, whereby he becomes 'dependent' on things and through them on the master who owns these things. The decisive insight is that domination and servitude only become possible through a particular form of the labour process: in the labour process existence for the servant becomes the 'chain from which he cannot abstract in the struggle'; the labour process is the cause and the safeguard of his 'dependence', just as, conversely, it is the cause and safeguard of the independence of the master.

Hegel's analysis of domination and servitude not only contains the justification of the authority of domination in the sphere of the social struggle: it also provides the *dialectic* of this authority. The immanent development of the relationship between domination and servitude not only leads to the recognition of servitude as the real 'truth' of domination, but also to the servant's own insight into the lord's real power and thus into its (possible) supersession; it is shown that the authority of the lord is, in the last analysis, dependent on the servant, who believes in it and sustains it.

Only through the labour performed in servitude does domination become real as a recognized power over the realm in which things are at its disposal. 'The truth of independent consciousness is accordingly the consciousness of the bondsman. . . . But just as domination shows its essential nature to be the reverse of what it wants to be, so, too, servitude will, when completed, pass into the opposite of what it immediately is. Being a consciousness repressed into itself, it will enter into itself, and transform itself into true independence.'[47] 'Fear' and 'service' (discipline and obedience), the features of the most extreme powerlessness and dependence, themselves become the productive forces which drive servitude out of its state of dependence. In the fear of 'absolute power' the consciousness of the servant is thrown back

47. Op. cit., p. 237.

on the 'simple essence of consciousness of himself', on his pure being for himself. And fear of the master becomes the 'beginning of wisdom': it forces the servant into the labour-process, in which his real power will reveal itself and in which he will come 'to himself'. Through the servant's work the immediate form of things is superseded by the only form in which they can be enjoyed and used. In the labour-process the 'subordinate consciousness' puts itself 'as such into the element of permanence; and thereby becomes for itself something existing for itself'. The form which it has given things, although it is 'put out' into the world of objects, does not become something alien or other: it is the manner of existence of its 'truth'; 'thus through this rediscovery of itself through itself, it acquires its *own meaning*, precisely in labour, where there seemed to be merely an *alien* meaning'.[48] And the real lever of further development, the supersession of the domination-servitude relationship, is not the dominating but the 'serving consciousness' which has acquired its true form in the labour-process.

This analysis of the relationship of domination and servitude doubtless marks the profoundest breakthrough of German Idealism into the dimension in which the social existence of man is built up as an authoritative order of domination. It is not absolute reason but absolute force which stands at the beginning of the 'objective spirit': the 'life and death struggle' for the recognition of property, the constitution of domination through the enslavement of the subordinated person in the labour-process. It is a long road to the total justification of the state by the absolute truth of the 'concept' – a road which nevertheless remains the prisoner of its origin. The young Hegel knew this: 'The concept . . . carries with it so much self-mistrust that it has to be validated by force, and only then does man submit to it.'[49]

48. Op. cit., p. 239.
49. *Hegel's Political Writings*, p. 242.

IV   Counter-revolution and Restoration

A. COUNTER-REVOLUTION

The theory of the counter-revolution emerged simultaneously with the French Revolution: Burke's *Reflections on the Revolution in France* appeared in 1790, Bonald's *Théorie du Pouvoir* and de Maistre's *Considérations sur la France* in 1796. Gentz, Friedrich Schlegel and Adam Müller undertook the propagation of their theories in Germany, and a straight line of social and ideological development leads from them to Friedrich J. Stahl's theory, elaborated under the Restoration in Germany. In the counter-revolution's philosophy of the state and society the theory of authority which subsequently becomes increasingly predominant is worked out for the first time – a consciously irrationalist and traditionalist theory. While the French use it clearly and trenchantly, mostly cynically, as a brilliant weapon in the political and social struggle, in the Germans it appears in an almost complete isolation from its actual basis; in the following we shall concentrate predominantly on its original form.

The theory of the counter-revolution initially fought for the feudal and clerical groups against the bourgeoisie as bearer of the revolution. In its long history it undergoes a decisive change of function: it is ultimately adapted for use by the ruling strata of the bourgeoisie. The bourgeoisie changes from object to subject of the theory. It is the finest example in modern times of the justification and defence of a threatened social order. The change of function of the theory accompanies the change in the history of the bourgeoisie from the struggle of a rising class against the remnants of a social organization which has become a fetter on it, to the absolute domination of a few privileged strata against the onslaught of all progressive forces; it also accompanies the alienation of the bourgeoisie from all the values which it had proclaimed at the time of its rise. It becomes clear precisely from the theory of the counter-revolution, in particular with regard to the problem of authority, how strong were the progressive tendencies in the bourgeois philosophy of the state and society.

This already emerges from a basic thesis common to the whole theory of counter-revolution,[1] which is directed against the bourgeois construction of state and society out of the rational will of man. If, in the face of this, state and society are now viewed, indirectly or directly, as divine institutions whose authority beyond this is derived either from its mere existence or mere permanence, or from a mystical *âme nationale* (de Maistre), this signifies the elevation of the existing system of domination above any possibility of justification *vis-à-vis* the insight and needs of individuals. The authoritative order embracing state and society is at once the 'divine and natural' order of things. 'Society is not the work of man, but the immediate result of the will of the Creator, who willed it that man should be what he has been always and everywhere.'[2]

Far from being able to constitute a state and a society by his own power, man can only 'retard the success of the efforts' made by a society in order to arrive at its 'natural constitution'. The political and religious constitutions of society 'result from the nature of human beings': 'they could not be anything other than they are, without colliding violently with the nature of the beings who compose each society'.[3] It is not the business of men to give society its constitution:[4] social organization can never be the subject of rational and deliberate human planning. That is the counter-attack not only against all bourgeois 'counter theories' (Rousseau's *Contrat social* is the initial target of the attack of counter-revolutionary theory), but also against any connection between state and society and the categories of 'reason': Hegel's

1. Carl Schmitt, *Politische Romantik*, 2nd edn, Munich, 1925, p. 153.
2. de Maistre, *Oeuvres complètes*, Lyon, 1891-2, vol. I, p. 317.
3. Bonald, *Oeuvres complètes* (ed. Migne), Paris, 1864, vol. I, pp. 121ff.
4. Ibid., p. 123.

theory of the state, too, is later attacked by the theory of the Restoration in the context of this main idea.

The civil constitution of the peoples is never 'the result of a discussion';[5] instead of this, God has given the people their government in two ways: he either leaves it to no one, 'as insensibly as a plant', or he uses 'rare men', 'truly chosen men to whom he entrusts his powers'.[6] The main motifs of the counter-revolutionary theory of authority are here united: the (theological-) *naturalist* and *personalist* justification of authority. A decisive tendency of the bourgeois theory of authority was the separation of office and person, the detachment of authority from its current personal bearer: basically it is not the (fortuitous) person who could justify the authority of the office but an order and legality which is somehow objective. That is now changed. Government becomes a charisma which is given by God to the current governing person as such and this charisma radiates out from the person of the ruler to the whole political and social order which culminates in him. This order is essentially personal and 'by nature' is centred on a single, indivisible personality: the monarch. 'In a situation where all men, having equal wills and unequal powers, necessarily wish to attain mastery, it is necessary that one man should be the master, otherwise all men would destroy each other.'[7]

This leads on the one hand to the irrational establishment of authority as an absolute: to the doctrine of the 'infallibility of the sovereign', and on the other to the total rejection of any attempt to change the prevailing rule of authority: to traditionalism. 'All possible sovereignties necessarily act in an infallible manner; for all government is absolute.'[8] Sovereignty is unconditionally 'valid', independently of its performance, its suitability, or its success; the ruler rules, because he possesses the 'royal spirit'. This is most clearly expressed in de Maistre's formula: 'It is generally believed that a family is royal because it reigns; on the contrary, it reigns because it is royal.'[9] (The German philosophy

5. Ibid., p. 346.   6. Ibid., p. 344.   7. Bonald, op. cit., p. 151.
8. de Maistre, op. cit., vol. II, p. 2; cf. vol. I, p. 417.   9. Ibid., vol. II, p. 421.

of the Restoration then disguised this clear-cut doctrine: C. L. von Haller endeavoured to show, with pages of argument, that in all areas of political and social life the rulers 'according to a universal law of nature', are also the most worthy.[10]

What, then, is the basis for the social life-process taking place in an order in which by far the greatest part of the people are subordinated to the unconditional domination of a few charismatically gifted persons? The divine order is at the same time the 'natural' order in the state of concupiscence, and the natural order is necessarily an order of classes: 'In all societies, consisting of different classes, certain classes must necessarily be uppermost. The apostles of equality therefore only change and pervert the natural order of things.'[11] 'Man, in his quality of being at once moral and corrupted, pure in his understanding and perverse in his wishes, must necessarily be subject to government.'[12] This appeal to the 'nature of man' leads back to the particular *anthropology* which underlies the theory of the counter-revolution as its most essential component.

It is an image of man which is drawn in terms of hate and contempt, but also of worldly wisdom and power: man who has fallen from God is an evil, cowardly, clumsy, half-blind animal which, if left on its own, only brings about dirt and disorder, which basically desires to be ruled and led and for which total dependence is ultimately the best thing. 'Sovereignty' originated with society itself: 'society, and sovereignty were born together'.[13] Whoever really knows the 'sad nature' of man knows also that 'man in general, if he is left to himself, is too wicked to be free'.[14] The natural wickedness of man corresponds to his natural weakness: the theory of the counter-revolution sanctions the total dependence of men on a few 'sovereigns' by engaging in a total defamation of human reason. 'Human reason, if we rely

10. C. L. von Haller, *Restauration der Staatswissenschaft*, Winterthur, 1820, vol. I, pp. 355ff.

11. Burke, *Reflections on the Revolution in France*, ed. H. P. Adams, London, 1927, p. 50.

12. de Maistre, op. cit., vol. II, p. 167.

13. Ibid., vol. I, p. 323.   14. Ibid., vol. II, p. 339.

solely on its innate powers, is nothing but a beast, and all its strength is reduced to the power of destruction.'[15] It is 'as much a nullity for the happiness of states as for that of the individual'. All great institutions derive their origin and their preservation from elsewhere; 'human reason . . . only involves itself in them in order to pervert and destroy them'.[16]

A similar tendency towards the devaluation of reason was already discernible in Luther and there too it was a part of his justification of worldly authorities. Here, however, in the theory of counter-revolution, every quietistic eschatological feature is obliterated: anti-rationalism is consciously wielded as an instrument in the class struggle, as an effective means of domination over the 'mass'; it has an explicitly political and activist character. One need only read the classic chapter: 'How will the counter-revolution be achieved, if it happens?', in de Maistre's *Considérations sur la France*.[17] And the most important element of this theory of domination over the masses is the theory of the social importance of *authority*.

'Men never respect what they have done':[18] this sentence indicates the basic motif. Since respect for the *status quo* is the psychological basis for the social order of domination, but this attitude is necessarily lacking in relation to works done by purely human might (what I have made, I can also destroy), state and society must be presented as something exceeding all human power: 'Every Constitution . . . is a creation in the full meaning of the expression, and every creation goes beyond the powers of man.'[19] The principle which upholds state and society is not the truth as arrived at through human insight, but faith: prejudice, superstition, religion and tradition are celebrated as the essential social virtues of man. Burke sings a hymn in praise of prejudice: 'Prejudice is of ready application in the hour of emergency. . . . It previously engages the mind in a steady course of wisdom and virtue. . . . Prejudice renders a man's virtue his habit. . . . Through just prejudice, his duty becomes part of his nature.'[20] De Maistre

15. Ibid., vol. I, p. 735.   16. Ibid., p. 367.   17. Ibid., vol. I, pp. 113ff.
18. Ibid., p. 353.   19. Ibid., p. 373.   20. Burke, op. cit., p. 90.

is even clearer: for man 'there is nothing so important . . . as prejudices'; they are 'the real elements of his happiness, and the watchdogs of empire'; without them there is 'neither religion, nor morality, nor government'. And he gives this instruction for the maintenance of every religious and political 'association': 'In order to conduct himself properly, man needs not problems but beliefs. His cradle should be surrounded with dogmas; and when his reason awakens, he should find all his opinions ready-made, at least on everything which relates to his behaviour.'[21] The true legislators knew why they intertwined religion and politics, 'so that the citizens are believers whose loyalty is exalted to the level of faith, and whose obedience is exalted to the level of enthusiasm and fanaticism'.[22]

The second form of domination over the masses as unquestioned subordination of 'individual reason' to universal prejudices is 'patriotism': 'the absolute and general reign of national dogmas, that is to say, useful prejudices.' The government is a 'true religion' which has its dogmas, mysteries and priests. 'Man's first need is that his dawning reason should be curbed beneath this double yoke, that it should obliterate itself, lose itself in the national reason.'[23] This conception of 'national soul' (*âme nationale*) and 'national reason' (*raison nationale*) here appears as an authority-producing factor in an anti-rationalist theory of domination over the masses; this is clearly very different from Hegel's concept of the people's spirit which, as the fulfilment of subjective and objective reason, was still linked with the rational will of individuals: the anti-bourgeois theory of the counter-revolution does not coincide with the philosophy of state originating in the rising bourgeoisie, even where their respective concepts have the most affinity. In the latter philosophy, the 'generality' into which the freedom of the individual was incorporated was meant at least in theory to fulfil the values and needs of the individuals in their 'superseded' form; the theory of the

21. de Maistre, op. cit., vol. I, p. 375.
22. Op. cit., p. 361. Burke calls religion 'the basis of civil society' (op. cit., p. 93).
23. Ibid., p. 376.

counter-revolution simply places the generality above all such values and needs. It stands above all human reason, beyond criticism and insight; for the individual it does not signify fulfilment but 'abnegation', 'annihilation'. The generality now stands in a negative relationship to the rational voluntariness of the individual: it simply demands his subordination. The apologia for religion and patriotism as the basis of society thus directly becomes the apologia of subordination and of an authority rising above all insight. After de Maistre has celebrated 'faith and patriotism' as the great 'healers of this world', he continues: 'they only know two words: *submission* and *faith*; with these two levers they raise the universe; their very errors are sublime.'[24]

If the social order is elevated as something divine and natural above the rational will and the plan-making insight of individuals, and if its authority is constantly held beyond the reach of critical insight by the psychological levers of religion, patriotism, tradition, prejudice, etc., this is meant to prevent the will of the 'mass of the people' from drawing conclusions from their perceptions, and undertaking to destroy an order of which they already know the origin and effect. This is not an interpretation, but the literal meaning of the texts of de Maistre and others. We quote the main passage from de Maistre's *Étude sur la Souveraineté* here, because in a few lines it indicates the arguments behind this whole theory of authority: 'To put it briefly, the mass of the people has absolutely no part in any political creation. It only respects the government itself because the government is not its own work. This feeling is engraved deeply into its heart. *It bends beneath the sovereign power because it feels that this is something sacred which it can neither create nor destroy.* If it succeeds in extinguishing from itself this preservative sentiment, owing to corruption and traitorous suggestions, if it has the misfortune to believe it is called, *en masse*, to reform the state, all is lost. This is why it is of infinite importance, even in the free states, that the men who govern should be separated from the mass of the people by the personal factor which results from birth and wealth: *for if the*

24. Ibid., p. 377.

opinion of the people does not place a barrier between itself and authority, if the power is not out of its reach, if the crowd who are governed can believe they are the equal of the small numbers who actually govern, *government no longer exists*: thus the aristocracy is the sovereign, or the ruling authority by its essence; and the principle of the French Revolution is in head-on conflict with the eternal laws of nature.'[25] The derivation of the decisive social relationships from authority is a central feature of the theory of the counter-revolution. Bonald endeavours to show that language, the first medium of socialization, is only received by the individual through authoritative communication.[26] The same goes for the law, science, art, methods of work, etc. 'Thus the initial means of all understanding is the word accepted on faith and without examination, and the initial means of education is authority.'[27] And, consistently with this, he defines the relationship between authority and reason in such a way that 'authority forms man's reason, by enlightening his spirit with the knowledge of the truth; authority placed the seeds of civilization in society . . .'.[28] The 'people' in particular, that is 'those whose purely mechanical and repetitious occupations keep them in a habitual state of childhood', are counted, along with women and children, among that class of people who because of their natural 'weakness' do not actively belong to society at all, but have to be protected by it. 'The people's reason must consist of *feelings*: we have to direct them, and form their *heart* rather than their *intellect*.' They have, then, to be kept in the state of weakness which is theirs by nature: reading and writing have to do with neither their physical nor moral happiness, are, in fact, not even in their interest.[29]

When authority is thus referred to as the 'seed of civilization', Bonald does not have in mind its 'domesticating' function in the sense of the regulation of the production process or the disposition of social labour with a view to the greatest possible exploitation of productive forces, but its *power of conservation and preservation*. The theory of the counter-revolution creates modern *tradi-*

25. Ibid., pp. 354ff. (my italics).   26. Bonald, op. cit., p. 1212.
27. Ibid., p. 1175.   28. Ibid., p. 1199.   29. Ibid., p. 747.

tionalism as a rescue operation for the endangered social order. The 'discovery of history' as the 'supreme master of politics', played off against the revolution 'without a history' has a purely reactionary character from this point onwards and right up to Moeller van den Bruck and 'existential philosophy': the historical, without regard for its material content, becomes an absolute force, which unconditionally subordinates man to the *status quo* as something which has always been and always will be; it even serves to 'destroy the category of time'.[30] History is only the preservation and handing on of what has existed in the past: 'every important and really constitutional institution never establishes anything new; it does nothing but proclaim and defend rights anterior to it.'[31] The 'new' is already in itself a sin against history. The binding and crippling power of such an attitude, if – as all the theorists of counter-revolution demand – it is impressed on the people from the cradle upwards through public and private education, is clearly recognized. For Burke, 'to be attached to the subdivision, to love the little platoon we belong to in society, is the first principle (the germ, as it were) of public affections'.[32]

But all this does not yet adequately describe the function of this irrationalist theory of authority. Its whole emotional effect was derived from its contemporary struggle against the French Revolution, in which (in Gentz's words) it saw the 'ultimate crime'. The divine and natural sanction of the social system of domination applies also, and to no small extent, to the inequality of property relations, and authority is to a considerable degree the authority of property. De Maistre gave this away by unquestioningly equating 'birth' and 'wealth', and Burke made the point openly: 'As property is sluggish, inert, and timid, it never can be safe from the invasions of ability, unless it be, out of all proportion, predominant in the representation. It must be represented too in great masses of accumulation, or it is not rightly protected. The characteristic essence of property formed out of the combined principles of its acquisition and conservation, is to be *unequal*.

30. H. J. Laski, *Authority in the Modern State*, New Haven, 1927, p. 127.
31. Bonald, op. cit., p. 373.   32. Burke, op. cit., p. 47.

The great masses therefore which excite envy, and tempt rapacity, must be put out of the possibility of danger. Then they form a natural rampart about the lesser properties in all their gradations.'[33] The decisive place of the family within the social system of authority also comes into view in the property context: 'The power of perpetuating our property in our families is one of the most valuable and interesting circumstances belonging to it, and that which tends the most to the perpetuation of society itself.'[34]

The idea of the inheritance of property is one of the most effective factors through which the family is tied to the order of state and society which protects it, and the individual is tied to the family; however, this is not the only reason why the family becomes a matter of life and death to the state. Authoritarian traditionalism knows very well that it is precisely in the family that the 'dogmas and prejudices' which it proposes as the basis of society are originally handed down: 'we know the morality that we have received from our fathers as an ensemble of dogmas or useful prejudices adopted by the national reason.'[35] The family is the basic image of all social domination, and although de Maistre does not wish to assert any 'exact parity between the authority of the father and the authority of the sovereign' he does say that 'the first man was the king of his children'.[36] Burke ascribes the stability of the English constitution to the fact that 'we took the fundamental laws into the womb of our families'; the authoritarian family becomes one of the key bulwarks against revolution: 'Always acting as if in the presence of canonized forefathers, the spirit of freedom . . . is tempered with an awful gravity.'[37] And to the ideal constitution of the family is added the genetic: Bonald claims that the 'political society' arose out of the struggle between 'proprietary families'.[38]

---

33. Op. cit., p. 52.
34. Loc. cit.
35. de Maistre, op. cit., vol. I, p. 400.
36. Ibid. p. 323.
37. Burke, op. cit., p. 34.
38. Bonald, op. cit., p. 1242.

B. RESTORATION

The spread of the theory of the counter-revolution in Germany and its transformation into the theory of the restoration took shape in two broad currents of thought: the first centred around 'political romanticism', beginning with Gentz's translation of Burke (1793) and reaching its peak in the time before and during the Vienna Congress (Friedrich Schlegel, Adam Müller, Baader, Görres); the second, the Restoration's theory of the state, was consolidated in Stahl's *Rechtsphilosophie* (the first edition of this book appeared at the time of the July Revolution, and it was given its final form in 1854). Von Haller's *Restauration der Staatswissenschaft* (from 1816 to 1834) represents, as it were, a link between the two currents. While, in France, the bourgeoisie was fighting to defeat the counter-revolution of the feudal aristocracy and after the July revolution consolidating its own political domination, the weaker economic development of the bourgeoisie in Germany led nowhere to real political power. The German theory of counter-revolution thus lacks all immediacy, trenchancy and aggressiveness; it isn't fighting against a revolution at all; the actual social antagonisms appear only fragmentarily through endless mediations. There is not a single decisive motif, for our context, which was not already present in the French theory of counter-revolution. But the situation has changed at the end of this process of development: the feudal monarchies are faced with the revolution. Stahl's theory of the authoritarian-theocratic state now becomes a welcome weapon in the open struggle.

The preface to the third edition (the second edition appeared a year after the revolt of the Silesian weavers and three years before the March Revolution in Germany) appeals to philosophy to come quickly to the aid of the threatened authorities in state and society; it is a convincing document on the justificatory and conservative function of philosophy in Germany.

'For one and a half centuries philosophy has not based government, marriage and property on God's order and providence but on the will of men and their contract, and the peoples were merely following its teachings when they raised themselves above their governments and all historically ordained orders and ultimately above rightfully existing property.'[1] This reproach is aimed not only at Rousseau and Kant, but also at Hegel, who may have proclaimed the 'sanctity of the concept' in place of the sovereignty of the will, but then 'who is afraid of this concept and who respects it?' It is precisely fear and respect which matter: philosophy must implant and sustain 'guilty obedience to authority'. Stahl exclaims anxiously: 'Should one leave the question "what is property?" only to the Proudhons?'[2] Philosophy should take over the great task of 'nurturing respect for all orders and governments which God has set over men, and for all conditions and laws, which have come into being in an orderly way under his directions'.[3]

Stahl's system (in its basis, not in its fully elaborated form, which reveals many concessions to bourgeois–liberal tendencies) is the first purely authoritarian German philosophy of the state, in so far as the social relations of people and the meaning and purpose of the political organization of society are from first to last directed towards the preservation and strengthening of an unassailable authority. The 'norms of civil order' are not taken from the real needs of the people, or from the general wish for the constitution of a true 'generality', or from a recognition of the progress of historical 'reason', but from the conception of a 'moral realm' whose cardinal feature is 'the necessity for an authority completely elevated above the people', that is, the necessity 'for a claim to obedience and respect which applies not only to the law but to a real power outside them – the government (state power)'. And since none of the theories which take the rational will of men as their starting point can ever arrive at an 'absolutely superior real authority', they are all, 'in their innermost foundations, revolutionary'.[4]

In the closer definition of this authority, the same tendencies converge as those already played off by the theory of the counter-

1. Stahl, *Rechtsphilosophie*, 3rd edn, Heidelberg, 1854–6, vol. II, 1, p. x.
2. Ibid., p. xvii.   3. Ibid., p. xxii.   4. Ibid., vol. II, 2, pp. 3ff.

revolution against the characteristics of the bourgeois social order; we have summarized them under the heading of the irrational personalization and the traditionalist stabilization of the existing (feudal-aristocratic) system of domination. The right of functioning authorities is freed from any justification through success and performance, and they are thus elevated charismatically above any control by society. To the removal of authority's material-objective character, through its fixation on the person who is 'gifted' with it, there corresponds on the other hand an (irrational) assertion of independence on the part of the state apparatus controlled by the persons in authority. Since the charismatic sanction can only consecrate a person and not an apparatus, the state as such must become a person: an independent 'organism', outlasting social changes, which, directly ordained by God, has its life outside the realm of individual and general aspirations. The irrational personalization of authority is transformed here, in Stahl's state-absolutism, into the most extreme form of reification: into the authority of the state seen as the supreme thing-in-itself. 'As the institution for the control of the entire condition of the human community the state is the *one, supreme and sovereign power* on earth. People and their aspirations, other institutions and communities, even the church, as far as its external existence is concerned, are subordinated to it. It judges them, without being judged by them or being able to be called to account by them, for there is no authority and no judge above it.'⁵ This reified state-absolutism was alien to the French counter-revolution: their image of domination was based too much on the nobility's personal pride and hate and personal contempt for the mass of the people to permit such a depersonalization. The independence of the state apparatus leaves open the possibility of replacing the bearers of authority, while retaining the supporting relations of production; the state philosophy of restoration leaves room for a compromise with the advancing bourgeoisie. There is a further indication of this in the connection between the absolute state and the 'soul of the people', with the people as an 'originally

5. Ibid., vol. II, 2, pp. 154f.

given unity'[6] in whose consciousness the state has its roots. This is no longer the 'national soul' of the counter-revolution, which was ultimately nothing more than an amalgamation of 'useful prejudices'. Stahl's concept of the soul of the people and of the people already clearly reflects a real participation by the dominated classes in their domination: the 'natural, organic community of the people' is meant to replace the social generality arising from the rational will of individuals which the bourgeois philosophy of the state had up till then required.

But the pure irrationality of the state-authority emerges again and again through the layer of ethical and organicist concepts which conceals it; this kind of authority can only demand obedience but cannot give a reason for it. The prestige of the state 'rests on its mere existence as such. It is an immanent, original prestige, and the subjects thus have the immediate duty to obey. . . . This obedience is not voluntary or dependent on consent but necessary; it is similar to one's duty towards one's parents. . . .'[7] Love and justice now acquire their real meaning as sanctions of the existing social system of authority. Love is based on obedience, which is 'the first and indispensable moral motive', and without which all love is merely 'pathological'.[8] And justice is defined as the 'inviolability of a given order . . . without regard to its content'.[9]

The ideological function of the law is proclaimed here with naïve openness. The organicists were aware why they placed such value on the traditional 'constancy' of the law: 'Through such constancy of the law the original simplicity of the people's consciousness is preserved, so that what is existing law is taken as just, and what is just as existing. Its effect is that law in itself is not known in any form other than the form of the law of the fatherland. . . . Hence the existing law is regarded as, by and large, what is necessary, and cannot be otherwise.'[10] Law and positive law become 'equivalent concepts'; there is no natural law or rational law which could be played off against the positive law.[11]

6. Ibid., pp. 234, 241.   7. Ibid., pp. 179ff.   8. Ibid., vol. II, 1, pp. 106ff.
9. Ibid., p. 163.   10. Ibid., p. 227.   11. Ibid., pp. 221ff.

The law regulates a social organization which is based on the triple pillars of 'the protection of the person', 'assets' and 'the family'.[12] It is the characteristic trinity which we have already met in bourgeois theory. Stahl too sees property as an 'original right of the personality', as the 'material for the revelation of the individuality of the man',[13] and lets the 'moral conservation' of property take place through the family; he rejects any attempt to base property on the will of men and traces it back directly to the providence of God. The social theory of the restoration is already confronted with a developed socialist theory: Stahl polemicizes against Considérant, Fourier and Proudhon. He knows that 'if the providence of God is not recognized as the legal basis of all property', there will be no reason for a private right to what is a means for the enjoyment and sustenance of all. 'Communism is thus correct as opposed to the philosophy of law from Grotius to Hegel, which bases property merely and finally on the will of man, and would be right as opposed to the present society if society itself would be prepared to free itself from God.'[14] Hence the necessity for a return from the human to the divine institution. But for the actual state of property relations an actual tribunal which can deputize for God is after all necessary. And here the anti-bourgeois character of the restoration breaks through again: 'the beginning of property among the people . . . is not appropriation by individuals but allocation by the government.' The historical legal basis for the distribution of property must not, under any circumstances, be founded – as in the bourgeois theory – on individual success and individual achievement. Property is to be based '*not on personal initiative but on authority*, not something gained through struggle but something received'.[15] It is the monarchical, feudal structure of authority which is expressed here: the authority of property does not depend on the individual property-owner, or on the general law safeguarding the individual property-owners, but on an ultimate 'government' from which the

12. Ibid., p. 310. 13. Ibid., p. 351.
14. Ibid., p. 375.
15. Ibid., p. 360 (my italics).

individuals receive their property as a 'fief' – although this distribution then becomes 'irrevocable'.[16]

In the course of the feudal-traditionalist theory of property the importance of the *family* for the stabilization of the authoritarian state is also recognized and defined. Only because it serves 'the revelation of individuality and the care of the family' does 'the moral consecration of property' take place.[17] It was precisely this function of the family which Hegel had already heavily emphasized in the finished form of his system and it is found at the same time in Riehl's typically bourgeois theory of the family. The theories of the Feudal Restoration and the liberal bourgeoisie meet in the celebration of the family as the material and moral foundation of society: the authoritarian and constitutionalist theories unite on the common ground of the protection of the family and of the order of property.

'The family is the centre of human existence, the link between the individual and the communal life', for in being 'the satisfaction of the individual' it is at the same time the means through which the civil and religious community 'comes into being both physically and morally and intellectually (through upbringing)'.[18] The political and social organization of the feudal monarchy as described by Stahl is so authoritarian in its construction that education towards authority in the family does not have to be particularly urged. Instead of this there is an emphatic indication that the endurance and the stability of the existing class order is largely due to the restriction of inheritance to within the family, which over many generations implants an interest in the continuity of this order in the individual consciousness: '. . . In this succession of families and of the wills modelled on them, there lies *order* and *continuity*, for the whole human race. Through this, mankind possesses property throughout a succession of generations, thus uninterruptedly controlling it as a substratum of the consciousness and the will, and preserving the legal groupings of people drawn up with respect to this property and through them

16. Ibid.
17. Ibid., vol. II, 2, p. 93.   18. Ibid., vol. II, 1, p. 424.

the connection between the generations.'[19] With regard to the material basis of the family community, Stahl recognizes that it is only through property that the 'goods of the earth' can become the instrument of 'family ties and family life'.[20] And he states that although 'in its most important aspect' the educational power has as its sole aim the 'advancement of the child', in addition to this, 'since the whole relationship also serves the satisfaction of the parents, it equally involves a domination of the children for the parents' own benefit: i.e. disposal of their services and labour'.[21]

19. Ibid., pp. 500ff.   20. Ibid., p. 352.   21. Ibid., p. 487.

On the road from Luther to Hegel bourgeois philosophy had increasingly dealt with the authority relationship as a social relationship of domination. It had thereby moved essentially from the centre to the periphery: the fixed centre was the Christian (inner, transcendental) freedom of the person, and the social order only appeared as the external sphere of this freedom. With values apportioned in this manner it was not difficult to accommodate the fact that the external sphere was primarily a realm of servitude and unfreedom, for this did not, after all, affect 'actual' freedom. Liberation always referred only to the inner realm of freedom: it was a 'spiritual' process, through which man became what he had always been in actuality. Since internal freedom always remained the eternal presupposition, or *a priori*, of unfreedom, external unfreedom could never close this gap: it was eternalized along with its opposite pole.

Since the eighteenth century there has been no lack of movements within bourgeois philosophy which have protested against this conception. The French Enlightenment made the concern for worldly freedom and the worldly happiness of men into a subject of philosophy: its limits were the limits of social order, which it could not essentially transcend. The only possibility of overcoming this whole conception lay beyond this order.

Behind the bourgeois concept of freedom with its unification of inner freedom and outer unfreedom Marx saw the Christian 'cult of the abstract man'; Christian freedom did not affect the social praxis of the concrete man (it was rather the unconditional authority of the 'law' or the worldly government which ruled there), but its actual 'inner' being as distinguished from its external existence. Thus the sphere in which men produced and reproduced their life appeared as a sphere of actual unfreedom and antagonisms, in which men were only counted free and equal as 'men' or as 'persons' without regard to their material existence. This image corresponds to bourgeois society as a society of commodity producers, in which men do not confront each other as concrete individuals but as abstract buyers and sellers of commodities and in which 'private labour' is expressed as abstract 'equal human labour', measurable in abstract social labour-time.[1] And a decisive presupposition of this society is the freedom of labour, in which all the features of the Christian bourgeois concept of freedom are realized. Freedom from all worldly goods here means that the worker has become 'free and independent' of all the things which are necessary for the preservation of his life; freedom of man to himself here means that he can freely dispose of the only thing which he still possesses, his labour-power; he has to sell it in order to live.[2] As far as he can sell it, he relates to it as to his 'property'. Bourgeois philosophy had taught that the freedom of the person could only be realized in free property. In this reality of bourgeois society one's own person has itself become property which is offered for sale on the market.

This revealing irony exposes the double truth which underlies the bourgeois categories: what this society has made of man, and what can be made out of him. The ground is laid bare on which the lever of transforming praxis can be put into operation in the direction of both poles. According to Marx the cultural values as well as the physical and psychological powers of men have become commodities under the capitalist mode of production. The situation of the labour market is what directly determines the freedom of men and the possibilities of life, and is itself always dependent on the dynamics of society as a whole.

Bourgeois philosophy's formulation of the problem was thus inverted; the same thing happened with the doctrine of the two realms of freedom and necessity and the dialectical relationship between them. The sphere of material production is and remains a 'realm of necessity': a perpetual struggle with nature determined by 'need and external requirement' and dependent on the

1. Marx, *Capital*, vol. I, Moscow, 1954, p. 73. 2. Op. cit., p. 169.

'more or less abundant conditions of production' in which it takes place.[3] But the realm of necessity also has its freedom; admittedly not 'transcendental' freedom, which leaves necessity behind and is satisfied with an 'inner' process. Marx had already traced the concept of necessity back to its content by including in it the real distress of men and their struggle with nature for the preservation of their life; he now did the same with the concept of freedom. 'Freedom in this field can only consist in socialized man, the associated producers, rationally regulating their interchange with nature, and bringing it under their common control instead of being ruled by it as if by a blind force; and achieving this with the minimum expenditure of energy and under conditions which are the most favourable to them and the most worthy of their human nature.'[4] Here for the first time freedom is understood as a mode of real human praxis, as a task of conscious social organization. The worldly happiness of men has been included in its content under the heading of 'the most adequate and the most worthy' conditions of human nature: the supersession of 'external' distress and 'external' servitude belong to the sense of this concept of freedom.

And yet there is still a 'higher' freedom: a 'development of human forces' which is not spurred on by need and external expediency, but 'is an end itself'. It only begins 'beyond' the sphere of material production, which will 'always remain a realm of necessity'. But its prerequisite is the rational organization of society: 'The true realm of freedom' can 'only blossom forth with that realm of necessity as its base. . . . The shortening of the working day is its basic prerequisite.'[5]

'The shortening of the working day is its basic prerequisite.' This sentence points to the injustice committed over a centuries-long development and gathers together the suffering and yearning of generations. The achievement of freedom is now recognized as one of the purposes of the organization of the social labour process and the appropriate form or organization has been determined:

3. *Capital*, vol. III, Moscow, 1959, p. 799.
4. Op. cit., vol. III, p. 800.   5. Loc. cit.

with this we are shown the road from the realm of necessity to the realm of freedom, a freedom which, although it is still something Beyond, is no longer the transcendental Beyond which eternally precedes man, or the religious Beyond which is meant to supersede their distress, but the Beyond which men can create for themselves if they transform a social order which has become rotten. The complete inversion of the problem of freedom, through which the realm of freedom as a particular 'worldly' organization of society is now founded on the realm of necessity, is only one aspect of the general inversion, in which the material relations of production of society are understood as the basis of the whole political and cultural 'superstructure' and its corresponding forms of consciousness.

In this connection, Marx also deals with the social bearing of the *problem of authority*. He confronts authority as a relationship of dependence in the capitalist process of production. His analysis is therefore concerned less with authority as such than with authority as a factor within a given society's relations of production. Only if we contrast this specific authority with the forms of authority prevalent in other societies do the more general functions of authority become visible.

Authority is a manifestation of the relationship of domination and servitude as a social relationship of dependence. However, the relationship of domination and servitude, 'as it grows directly out of production itself', is determined by the 'specific economic form in which unpaid surplus labour is pumped out of the immediate producers'.[6] The specific form of the capitalist labour process determines the form of the authority relationships predominant in capitalist society. This labour process[7] requires the 'cooperation of many wage-labourers' first in cottage industry and later in the factory, and 'social or communal labour on a larger scale'. Such labour necessarily has to have a management which unites the individual activities, caused by the division of labour,

6. Op. cit., vol. III, p. 324.
7. The quotations in the following three paragraphs are from *Capital*, vol. I, pp. 330ff.

into a 'productive overall body': it mediates, supervises and leads. Since the means of production and the immediate conditions of production were in the possession of capital, this function of management necessarily fell to the capitalist: originally the 'command of capital over labour' appeared to be 'only a formal consequence of the fact that the worker works, not for himself, but for the capitalist and therefore under the capitalist'. In so far as the dominating authority of the capitalist is a 'direct requirement for the carrying out of the labour process' it is a real requisite of production: the capitalist's command on the field of production is as indispensable as the general's command on the field of battle.

But this is only one side of authority. The capitalist production process aims for the greatest possible production of surplus value, i.e. for the greatest possible exploitation of the labour power of the wage labourer. The greater their number grows and the more their resistance to their economic situation increases, the fiercer the pressure of the dominating authority of capital. 'The management of the capitalist is not only a function which springs from the social labour process and which particularly appertains to him, it is at the same time a function of the exploitation of a social labour process and thus dependent on the inevitable antagonisms between the exploiter and the raw material of his exploitation.' This is the second side of authority. And in this two-sidedness it now determines the specific form which the relationships of domination and dependence assume in capitalist society.

There now grows out of the dialectic of the labour relationship what Marx has called the 'despotic form' of capitalist management. It comes into being when, in the development of production, the following two functions of management directly coincide: the function which springs from the communal labour process, and the function which springs from the process of the realization of capital, that is, authority as a condition of production, and authority as exploitation. The 'office' of management is not the result of the material rational organization of the labour process, but appears as an adjunct to the ownership of the means of production: it becomes the prerogative of the capitalist. 'The capitalist is not a capitalist because he is an industrial commander but becomes an industrial commander because he is a capitalist.' The division of labour is reified and stabilized so that it becomes a 'natural' division between the disposition and the execution of labour; the 'labour of overall supervision hardens into its exclusive function'.

This whole process is constantly reproduced under the compulsion of economic necessity; it runs its course as it were behind the backs of the men who are subject to it. The authority which springs from economic power appears to them as the personal authority of the capitalist, as the 'power of an alien will, which submits its activity to his purposes'. Reification is transformed into a false personalization: whoever happens to be the manager of the labour process is always ready in possession of an authority which, properly speaking, could only emerge from the actual prior management of the labour process. The capitalist possesses and uses his authority *vis-à-vis* the workers essentially as the 'personification of capital; his personal authority *vis-à-vis* the workers' is only the 'personification of the conditions of labour *vis-à-vis* labour itself'.[8]

This analysis of the authority relationship as it has grown directly out of the production process also shows how the irrational personalization of authorities typical of the later period is anchored in the essence of the capitalist production process. It further shows that the existence side-by-side of an authoritarian and an anti-authoritarian attitude, which we have been able to follow right through bourgeois philosophy, similarly springs from the peculiar character of this process.[9]

While the division of labour within the workshop or factory uniformly subjects the cooperating workers to the unconditional authority of the capitalist and creates a purposeful and despotic form of management, the social division of labour itself, as a whole, is still left to an arbitrariness following no rules: 'chance and caprice have full play in distributing the producers and their

8. Op. cit., vol. III, p. 418.
9. What follows is based on *Capital*, vol. I, p. 355.

means of production, among the various branches of social labour.' Without a plan regulating the overall process of production, the independent commodity producers confront each other without 'acknowledging any authority other than that of competition, of the coercion exerted by the pressure of their mutual interests'. The more the anarchy which permeates the whole social process spreads, the more despotic will the authority of the capitalist over the immediate producers become in the labour process itself. In capitalist society 'anarchy in the social division of labour and despotism in that of the workshop mutually condition each other'. In a retrospective glance at pre-capitalist social organizations Marx differentiates between the anarchic-despotic authority structure just described and the authority relationships predominant in those earlier societies.[10] The relationship between the social and 'factory-type' division of labour is there exactly reversed: while the overall social division of labour is subjected to a 'planned and authoritative organization', the division of labour in the workplace is not at all developed or only undergoes a 'dwarf-like' development. Marx points to the example of the small Indian communities: the specialization of trade develops 'spontaneously' out of the given forces and conditions of production and crystallizes into a legal system, authoritatively and systematically regulating the community's relations of production; while within the individual trades each artisan works, although exactly according to tradition, 'independently and without acknowledging any authority'. Here too the law of the overall social division of labour has 'the irresistible authority of a law of nature'; but this 'law of nature' is comprehensible to the people there who are subject to it, and to a great extent it is a 'natural' law which regulates the reproduction of society according to the natural and historical conditions of production; whereas in capitalist society it is opaque and operates as an alien force resistant to the possibilities already available.

In summary Marx lays down the distinction between the

10. *Capital*, vol. I, pp. 357–9, and *The Poverty of Philosophy*, Moscow, 1966, p. 118.

distribution of authority in capitalist and in precapitalist societies as a 'general rule': 'the less authority presides over the division of labour inside society, the more the division of labour develops inside the workshop, and the more it is subjected to the authority of a single person. Thus authority in the workshop and authority in society, in relation to the division of labour, are in *inverse ratio* to each other.'[11]

The dialectical and two-sided character of the authority relationship is also the determining factor in the establishment of a *positive concept of authority*; this became a particularly central preoccupation in the debate with the anti-authoritarian anarchies of the followers of Bakunin. A small essay by Engels, *On the Principle of Authority*, summarizes the principal points of this discussion.[12]

In contrast to the undialectical rejection of all authority, emphasis is first laid on the dialectical character of the authority relationship: it is an 'absurdity' to present the principle of authority as absolutely bad and the principle of autonomy as absolutely good. There is a kind of authority which is inseparably linked with all 'organization', a kind of subordination, based on functional-rational assumptions, to genuine management and performance-labour discipline. Such functional authority is necessary in every social organization as a condition of production; it will also play an important role in a future society. Admittedly this society will only allow authority to exist within the bounds 'inevitably drawn by the relations of production'. The features of the authority structure determined by class society will disappear, in particular the function of exploitation and the political appropriation of 'management' in the capitalist system of domination. Public function will lose this political character and change into 'simple administrative functions'; those who fulfil these functions will watch over the social interests of the whole society.

11. *The Poverty of Philosophy*, p. 118.
12. Engels, *Von der Autorität*, Marx-Engels Werke, Berlin, 1960, vol. 18, pp. 305–308 (originally written in Italian for the *Almanacco Repubblicano per l'anno 1874*).

Engels holds up another decisive function of genuine authority as an objection against the anti-authoritarian: the role of leadership and the leading party in the revolution. 'A revolution is certainly the most authoritarian thing there is, an act in which one part of the population forces its will on the other with muskets, bayonets and cannons, which are all very authoritarian means.' Revolutionary subordination in one's own ranks and revolutionary authority towards the class enemy are necessary prerequisites in the struggle for the future organization of society.

This progressive function of authority was more closely defined by Lenin in the context of his struggle against 'economism'. The authority of rational leadership is separated off by Lenin from anarchism on the one hand and the theory of spontaneity on the other. The worship of the spontaneous mass movement which pursues its aim unaided, and the related disparagement of the initiative of the leaders signifies 'converting the working-class movement into an instrument of bourgeois democracy'.[13] The 'conscious element' is a decisive factor in the movement; to weaken it means to strengthen bourgeois and in particular petty-bourgeois influence. 'Class political consciousness can be brought to the workers only *from without*, that is, only from outside the economic struggle, from outside the sphere of relations between workers and employers.'[14] From the importance of the conscious element there emerges the necessity for a strict, centralist organization with a proven and schooled leadership at its head. Lenin claims that 'no revolutionary movement can endure without a stable organization of leaders maintaining continuity' and that 'the broader the popular mass drawn spontaneously into the struggle, which forms the basis of the movement and participates in it, the more urgent the need for such an organization'.[15]

In Marx the starting-point for the analysis of authority was the interest which a particular society had in subordinating people to a directing will within the material process of production and reproduction. In capitalist society this interest is first and last the

---

13. *What is to be Done?*, Moscow, 1969, p. 194.
14. Op. cit., p. 78.   15. Op. cit., p. 121.

interest of the ruling class, an interest growingly antagonistic to the interest of the great majority, even if – thanks to the double-edged character of the authority relationship in this case – to a certain (and increasingly problematic) degree, the interest of the whole of society was thereby served. The material root of the authority relationship described was the specific form of the capitalist production process: 'The immediate relationship of the owners of the conditions of production to the immediate producers.' But the social function of authority is by no means exhausted in this immediate relationship and its immediate consequences. Through numerous mediations it extends from this point to embrace the entire compass of human social organization. Marx followed the main directions of these mediations: he dealt with the problem under the most varied headings (state, law, tradition, history, etc.) and let it lead into the ultimate question of the reality of the social freedom of man. In the following we shall only point to some of the questions which are directly relevant to the problem of authority.

First we must remember that the 'domination-servitude relationship, as it grows directly out of production itself . . . in turn has a determining effect on the latter'.[16] It is one of those social relationships which, once they have come into being at a particular stage of the production process, build up a powerful resistance to the development of this process, harden into their acquired form and in this hardened form influence the material life process of society. This mechanism, by which an authoritative relationship of domination, originally made possible through the labour process, extends and stabilizes itself beyond its origins, this 'reification' of authority, occurs partly 'by itself', partly as the praxis of the ruling groups. The reification occurs by itself when the basis of an existing state of social production is constantly reproduced and assumes a regulated and ordered form (regulation and order are themselves an 'indispensable element in every mode of production'). It occurs as the praxis of the ruling group, because it is in their interest to 'sanctify as law' the existing state of

16. *Capital*, vol. III, p. 324.

affairs, in which they have risen to a position of domination. It is the authority of *tradition*, in which Marx here reveals the same double-edged quality as exists in the authority of the director of labour: the private appropriation of a social interest and its transformation into an instrument of economic and psychological domination.

Marx discovered the same double-edged duality, determined by the material relations of production in capitalist society, in those authority relationships which have the most 'general' character: in the political organization of society. Bourgeois philosophy had essentially understood the problem of social unfreedom as the problem of the unification of the individual and the generality (the supersession of individual freedom in the generality); Marx, in investigating this generality in a historical materialist way, shows its character as *appearance* in previous history and reveals the mechanism which turns the appearance into a real force.

What is the importance of the general in the social existence of men? Firstly, nothing other than the 'mutual interdependence of the individuals among whom the labour is divided',[17] their common neediness, their common reliance on the available productive forces and conditions of production. The general interest is the reproduction of the whole society under the best exploitation of the productive forces available, for the greatest possible happiness of the individuals. In every society in which labour is divided and appropriated according to class, in which the acquisition of surplus value occurs at the cost of the immediate producers, a contradiction necessarily appears between the general interest and the interest of the ruling class. And 'precisely from this contradiction between the particular and the general interest' the *state* assumes an apparently independent form. The process tending towards the independence and consolidation of the general as an alien and independent power, separated from the wishes and acts of the individuals, is one of the decisive elements in the universal reification which was already present in the

17. *The German Ideology*, Moscow, 1968, p. 44.

authority of the 'management' of labour. And here too the process is double-edged. On the one hand the ruling class, in order to justify its dominating position in the process of production, has to make the particular interest of its class seem valid as the general interest, 'that is expressed in ideal form, to give its ideas the form of universality, and represent them as the only rational, universally valid ones'.[18] Thus far, the general is merely a 'creation' of individuals who are defined as private people and the contradiction between the general and the private interest, like the independence of the general, is only an 'appearance',[19] which is produced again and again in history and destroyed again and again. On the other hand the independence of the apparently general is based on a very real power: the state in all its institutions as genuine force. The perpetually conflicting activity, the perpetual struggle between 'opposed particular interests' requires, if the reproduction of the anarchically producing society is to be safeguarded, a universal apparatus which is equipped with all the material and intellectual instruments of coercion: it 'makes practical intervention and control necessary through the illusory "general" interest in the form of the state'.[20]

The analysis of the concrete social character of the generality, of its nature as an appearance which is nevertheless real, now also leads to the critique of the bourgeois concept of freedom.

The personal freedom, which bourgeois society did in fact develop in contrast to the personal bondage of feudalism, is the expression of the free competition of commodity producers. Freedom of labour, freedom of movement, freedom of occupation, freedom of profit – all these varieties of bourgeois freedom express the 'accidental nature of the conditions of life', which the capitalist production process has brought forth in general competition and in the general struggle of individuals amongst each other.[21] Such freedom is merely fortuitous – in fact, the personality itself becomes something fortuitous and fortuitousness becomes a personality.[22] And what asserts itself in the overall

18. Op. cit., p. 62.   19. Op. cit., p. 272.
20. Op. cit., p. 46.   21. Op. cit., p. 95.   22. Op. cit., p. 421.

society in the form of this fortuitousness is only that anarchic form of its reproduction. It is on this and particularly on the transformation into wages of the value and the price of labour power (obscuring the real relationship) that 'all the illusions of freedom' in bourgeois society are based.[23] Its freedom is only the phenomenal form of general unfreedom, powerlessness in relation to the social production process, which for these people becomes a 'material force' by which they are ruled instead of ruling it. Freedom is only possible in the general community: that was the correct answer of bourgeois philosophy. But the general community which makes freedom possible is a quite particular form of organization of the whole society which can only be realized through the supersession of its bourgeois organization. The latter was an 'apparent' universality in which the unification of the individuals signified general unfreedom. In a genuine universality 'the individuals obtain their freedom in and through their association'.[24] 'In place of the old bourgeois society . . . we shall have an association in which the free development of each is the condition for the free development of all.'[25]

We have pursued the authority problem down to its most general formulations because only thus could we show that for Marx it is entirely a social problem, which can only be tackled by a particular social praxis at a particular stage of historical development. Marx's work is not a description of social conditions, but the theory of tendencies of social development. The supersession of capitalist by socialist society is an historical tendency which is itself at work in the given social situation. 'The "Idea" always disguised itself insofar as it differed from the "Interest".'[26] The decisive authority is not the idea (not even the idea of a just and free society), but *history*. Only in history can there originate the 'interest' which is needed by the idea for its realization.

The materialist analysis of the tendencies of the capitalist

---

23. *Capital*, vol. I, p. 540. 24. *The German Ideology*, p. 93.
25. *The Manifesto of the Communist Party*, in Marx-Engels, *Selected Works*, vol. I, Moscow, 1956, p. 109. 26. *The Holy Family*, Moscow, 1956, p. 109.

production process now also attacks an element of bourgeois theory which had been of decisive importance ever since Luther: the idea of the *family* as the moral foundation of the social system of domination. The concept of the family is an indifferent abstraction (which none the less makes good ideological sense as the perpetuation and generalization of a particular form of the family); it is the form of the patriarchal, monogamous, nuclear family which, in the long historical development beginning at a particular stage of the social life process, obtains objective status as an important element in this process.[27] Marx distinguishes the ideological appearance of the bourgeois family from its material reality; existing theories had so far put the two together.[28]

The reality of the bourgeois family is determined, like all forms of life under capitalism, by the character of the commodity economy; as a 'property' with its specific costs and expenses, profit and surplus value, it is entered into the general account. Economic interests govern not only the choice of partner (mostly prescribed by the father) but also the production and upbringing of the children. Like the physiological functions, the spiritual values are also tied to the economic interests; in their accustomed and comfortable form they govern day to day cohabitation. 'The bourgeoisie historically gives the family the character of the bourgeois family, in which boredom and money are the binding link.'[29] On this basis there now appear the phenomena characterized by Marx as the apparent dissolution of the bourgeois family by the bourgeoisie itself: the breaking of monogamy through 'secret adultery', the hidden 'community of married women', prostitution, etc. While on the one hand the bourgeoisie has 'torn away from the family its sentimental veil, and has reduced the family relation to a mere money relation',[28] this 'dirty existence' of the family has its counterpart on the other hand in the 'holy concept of it in official phraseology and universal

27. We shall not here go into the historical-genetic theory of the family developed by Engels.
28. On this distinction, see especially *The German Ideology*, pp. 195–8.
29. Op. cit., p. 195.
30. *The Manifesto of the Communist Party*, p. 37.

hypocrisy'.[31] For the bourgeoisie has a vital interest in the continued existence of the family because marriage, property and the family are 'the practical basis on which the bourgeoisie has erected its domination, and because in their bourgeois form they are conditions which make the bourgeois a bourgeois'.[32] This is the materialist formulation of the relationship, idealized after Marx by bourgeois theory, in which the family of private property-owners is made into the moral foundation of society. The bourgeois family continues to exist because its existence 'has been made necessary by its connection with the mode of production that exists independently of the will of bourgeois society'.[33] While its dissolution is only an apparent one, this mode of production leads on the opposite side – in the proletariat – to a real dissolution of the family. Marx has portrayed the terrible destruction of the proletarian family by large-scale industry from the middle of the nineteenth century:[34] the exploitation of the labour of women and children dissolved the economic base of the old family; to the increased general exploitation was added the as it were additional exploitation of wife and children by the father, who was driven to the selling of both.

If capitalism thus actually perverted all apparently 'eternal' and 'natural' family relationships, it was nevertheless precisely through this that it made visible the social determination of the existing form of the family and the way to overcome it. Large-scale industry 'by assigning as it does an important part in the process of production, outside the domestic sphere, to women, to young persons, and to children of both sexes, creates a new economical foundation for a higher form of the family and of the relations between the sexes'.[35] The functions fulfilled by the bourgeois family will be freed from their connections with the characteristics of the capitalist production process: authority will be separated from the interest of exploitation, the education of

31. *The German Ideology*, p. 195.
32. Loc. cit.
33. Op. cit., p. 196.
34. *Capital*, vol. I, pp. 480ff.
35. Op. cit., p. 490.

children from the interest of private property. This will result in the destruction of the two bases of marriage so far: 'the dependence of the woman on the man and of the children on their parents through private property.'[36]

36. *Marx-Engels Gesamtausgabe*, part 1, vol. VI, p. 519.

The transformation of
the bourgeois theory of authority
into the theory
of the totalitarian state (Sorel and Pareto)

A good deal of the history of bourgeois society is reflected in the bourgeois theory of authority. When the bourgeoisie had won political and economic domination in Western and Central Europe the contradictions within the society it organized were obvious. As the ruling class the bourgeoisie could hardly retain its interest in the theory with which it had been linked as a rising class and which was in crying contradiction with the present. This is why the actual bourgeois theory of society is to be found only before the real domination of the bourgeoisie, and the theory of the dominant bourgeoisie is no longer bourgeois theory. Comte was the last man in France, Hegel was the last man in Germany to discuss the problems of social organization within a comprehensive theory as tasks for rational human praxis.

Problems of the organization of state and society, once they have broken away from the supporting foundation of the comprehensive theory, fall to the business of the specialist discipline of *sociology*. A brief survey will be given elsewhere[1] of the forms assumed by the problem of authority in the various tendencies of bourgeois sociology. They are symptomatic of certain stages and streams within the development of society, but none offers a new interpretation of social domination and none consciously expresses a new overall social constellation.[2] The real bourgeois theory

1. Cf. my essay *Autorität und Familie in der deutschen Soziologie bis 1933*, Paris, 1936.

2. We omit here the theory of the 'Basel Circle', particularly of Nietzsche and Burckhardt, which contains decisive insights into the development of society. Their concrete social importance has not yet been recognized. They have had no effect up to the present: their current derivations stand in total contradiction to their actual content. Cf. references to this state of affairs in *Zeitschrift für Sozialforschung*, IV (1935), pp. 15ff.

continues in a weak and, as regards content, an increasingly thin line (the neo-Kantian philosophy of law); the more the liberal bourgeoisie transforms itself and goes over to anti-liberal forms of domination, the more abstract becomes the theory of the state (the theory of the formal legal state) which still clings to the liberalist foundations.

Only at the present time of preparation for world war do the elements of a new theory of social domination corresponding to a new overall situation come together. This theory has taken on a firm shape simultaneously with the abolition of democratic and parliamentarian forms of government in Central and Southern Europe. The bourgeoisie has retained its domination by retaining the leadership of the smallest, economically most powerful, groups. The total political apparatus is built up under the most severe economic crises. Social relationships of authority assume a new form. Theory as a whole attains a different significance: it is consciously 'politicized' and made into the weapon of the total authoritarian state.

The unity of bourgeois theory at this stage is *negative*: it rests exclusively on the united front against liberalism and Marxism. It is the enemy who prescribes the position of the theory. It has no ground of its own from which the totality of social phenomena could be understood. All its basic concepts are counter-concepts: it invents the 'organic' view of history in opposition to historical materialism, 'heroic realism' in opposition to liberal idealism, 'existentialist philosophy' in opposition to the rationalist social theory of the bourgeoisie, and the totally authoritarian 'Führer-staat' in opposition to the rational state. The material social content of the theory, i.e. the particular form of the relations of production, for the maintenance of which it functions, is obscured.

This determines a basic characteristic of the theory: its *formalism*. This may seem strange, since it is precisely material contents (like race, people, blood, earth) which are brought into the field against the formal rationalism of the old theory of state and society. But where these concepts are not yet in the forefront (as in Pareto) or represent a later disguise (as in Carl Schmitt) the formal character of the theory becomes obvious. We will illustrate this directly with reference to the concept of authority.

Seen from the previous stage of its development, the relationship of authority and domination is defined in such a way that authority is not seen as a function of domination, a means of dominating, etc., but as the basis of domination. Authority as power over voluntary recognition and over the voluntary subordination to the will and insight of the bearer of authority, is a 'quality' which certain people have 'by birth'. This seems at first sight to be merely a revival of the charismatic justification of authority; but this is not the case, for the charisma of authority is itself in turn 'justified' (without direct recourse to God). Its prerequisite is that the bearer of authority should belong to a given 'people' (*Volkstum*) or a given 'race': his authority rests on the genuine 'identity of origin' of the leader and the led.[3] This very broad biological basis makes it possible to extend charismatic authority at will to any number of people throughout all social groups. How can the hierarchy of authorities necessary for social domination within a total-authority system be built on such a formation, if social development has made every 'generally valid' rational and material criterion for the necessity of the required system of authority impossible?

After every possible rational and material content of authority has fallen away only its mere form remains: authority as such becomes the essential feature of the authoritarian state.[4] The absolute activity and the absolute decision of the leading men obtain a value independent of the social content of their acts and decisions. The absolute acceptance of their decision, the 'heroic' sacrifice of the led, becomes a value independent of insight into its social purpose. According to this theory society is not divided into rich and poor, nor into happy and miserable, nor into progressive and reactionary, but with the cancellation of all these material contradictions, into leaders and led. And the specific hierarchy of such an authority system hangs (since the merely

3. Carl Schmitt, *Staat, Bewegung, Volk*, Hamburg, 1933, p. 42.
4. Koellreutter, *Allgemeine Staatslehre*, Tübingen, 1933, p. 58.

biological identity of origin on human society does not create any hierarchical gradations) in thin air: the leading 'élites' can be changed at will according to the requirements of the power groups standing behind them.

The formalism of the authoritarian theory of the state is the thin veil which reveals more than it conceals of the actual constellation of power. It shows the distance which separates the new theory from the genuine bourgeois philosophy of state and society. Quite unjustifiably it invokes Hegel's idea of the 'organic' state, to which its anti-rationalism is in utter contradiction. And not only that: Hegel's philosophy is entirely 'material' in these dimensions. It measures the rationality of the state by the material progress of society and is, as one can imagine, unsuited for the defence of the total-authoritarian state. And those of its defenders who make the struggle against German Idealism a heart-felt test of 'heroic realism' are here guided by a more accurate instinct.[5]

We shall not go into the theory of the totally authoritarian state;[6] we shall merely deal briefly with the theory of Sorel and Pareto as the transition to the present-day conception of authority.

A. SOREL

In Sorel's work (from 1898, the year in which *L'avenir socialiste des syndicats* appeared) the changed social situation, which necessitates changed tactics in the social struggle, is announced for the first time in sociological literature. Sorel's anarcho-syndicalism, his myth of the eschatological general strike, and of the proletarian violence which will 'unalterably' destroy the bourgeois order, seem a long way from the theory of the authoritarian state.

5. E.g. Ernst Krieck, in his essays in the periodical *Volk im Werden*, 1933, and in his book *Nationalpolitische Erziehung*.
6. Some of the connections between the total-authoritarian theory of the state and the problem discussed here are presented in *Zeitschrift für Sozialforschung*, III (1934), pp. 161ff.; trans. J. J. Schapiro and printed in Marcuse, *Negations*, London, 1968, pp. 3–42.

Sorel's position and influence is ambiguous;[7] we shall not attempt a new categorization here. We shall merely seek to bring out a few features of his work which pave the way for the theory of the authoritarian state.

Sorel's work is a typical example of the transformation of an abstract anti-authoritarian attitude into reinforced authoritarianism. Sorel struggles against organized centralism under the guidance of the party leadership, against the political organization of the proletariat as a 'power formation'; he demands a 'loosened, federalized world of proletarian institutions and associations'; an 'acephalous' socialist movement.[8] This anti-authoritarian anarchism is closely tied to the freeing of socialism from its economic basis: to its transformation into a 'metaphysics of morals'.[9] Materialism is abandoned at one of its decisive points: 'Socialism as the promise of sensual happiness is destruction'[10] – a sentence which is not made less significant even by Sorel's attacks on the Idealists.

The failure to recognize the meaning of authority as a condition of all (even socialist) 'organization' is only an expression of the removal of the socialist base just referred to. Proletarian 'violence', which along with the myth of the general strike is engaged in the final struggle with the bourgeois order, is separated from its economic and social purpose; it becomes an authority in itself. If its criterion no longer lies in material rationality and greater happiness in the social life-process towards which this force is directed, then there is no rational explanation whatsoever as to why proletarian should be 'better' than bourgeois violence. In its effect, Sorel's work, with its strong attacks on soggy liberalism, the degeneration of parliament, the cowardly willingness to compromise, the pre-eminence of intellectuals, etc., could just as easily be taken as a call to the bourgeoisie openly to use the power which it clearly factually possesses: 'It is here that the role of violence in history appears to us as singularly great, for it can, in

7. M. Freund, *Georges Sorel*, Frankfurt (1932), contains a good compilation of the material.

8. Ibid., p. 105.   9. Ibid.   10. Ibid., p. 104.

an indirect manner, so operate on the middle class as to awaken them to a sense of their own class sentiment.'[11]

In a decisive context Sorel himself emphasized the central importance of authority in the revolutionary movement: in connection with the question, on the basis of what authority the workers would be kept to increased labour-discipline in the production process after the struggle had been won.[12] The authority problem here appears under the heading of revolutionary 'discipline': Sorel establishes a basic distinction between the 'discipline which imposes a general stoppage of work on the workers, and the discipline which can lead them to handle machinery with greater skill'. He separates this positive authority from any external coercion and seeks its basis in a new 'ethics of the producers', a free integration of the individual into the collective. The 'acephaly' of socialism is transformed into the theory of revolutionary 'élites': social revolution gives birth to new 'social authorities' which 'grow organically' out of social life and take over the disciplinary leadership of the production process. The élite as bearer of future 'social authority' is an élite of 'social merit': it consists of 'groups, which enjoy a moral hegemony, a correct feeling for tradition and in a rational manner care for the future'.[13]

Direct lines of development have been drawn from Sorel's concept of social élites to both the proletarian 'avant-garde' of Leninism and to the élite 'leaders' of Fascism. Freed from the connection with a clear economic base and elevated into the 'moral' sphere, the conception of the élite tends towards formalistic authoritarianism. We shall now examine this tendency and briefly look at the form which the concept of the élite assumed in Pareto's sociology.

11. *Reflections on Violence*, trans. T. E. Hulme and J. Roth, London, 1970, p. 90. Cf. the apology for violent and cunning capitalists, op. cit., pp. 86ff.
12. Op. cit., p. 237.   13. Freund, op. cit., p. 215.

B. PARETO

Pareto's concept of the élite is part of a rationalist-positivist social theory which for the most part constructs the social 'equilibrium', especially the stability of domination and being dominated, on irrational factors: on the functioning of certain psychological mechanisms and their derivations. This sociology has achieved the ideal state of a complete 'freedom from values': with overt cynicism it dispenses with any 'moral' standpoint at all towards social processes. But it also dispenses with any standpoint towards their material content. The economic matter of social production and reproduction is of no interest to it: it only describes what is meant to have occurred on a given material base in all times and in all places. Nevertheless there is no doubt here as to the social groups in whose interest its formalism functions.

Society, which is necessarily and by nature heterogeneous, falls for Pareto into two strata: 'a lower stratum, the *non-élite*, and a higher stratum, the *élite*, which is divided into two: (a) a governing *élite*; (b) a non-governing *élite*.'[1]

The ruling élite is constituted on the basis of the degree of 'capacity' through which the individual distinguishes himself in his 'profession'. The 'profession' itself is not immediately relevant. The great courtesan and the great capitalist, the great confidence trickster and the great general, the great poet and the great gambler in this manner belong to the superior class, the élite,[2] and, if they somehow succeed in obtaining influence on the ruling group, also the 'governing élite'. To get 'on top' and be able to stay 'on top' becomes the only criterion of the élite, where 'on top' is defined purely formally as opposed to 'below': as the power and disposal over other people and things (no matter in which areas and for which ends this power is used).

In this conception of the élite there are still strong *liberal*

---

1. V. Pareto, *The Mind and Society: A Treatise on General Sociology*, trans. A. Bongiorno and A. Livingstone, New York, 1935.
2. Op. cit., para. 2027.

elements: the elbow-room of the aspiring bourgeoisie, the pure 'ideology of success', the individual possibility for everyone of rising from every social position. These are reinforced even more by the theory of the 'circulation of élites': new and refreshing streams from the *lower class* penetrate the *higher class* which in its constitution otherwise becomes increasingly rigid or flabby: 'The governing class is restored ... by families which rise from the lower classes and bring with them the vigour and the proportions of residues necessary to keep it in power.'[3] Revolution, as a sudden and forceful replacement of one élite by another, is as it were only a disturbance in the normal circulation process.[4] It is a decisive feature of this theory that it replaces the material division of society into classes by a formal division, which itself in turn fluctuates, going diagonally through classes according to 'abilities' (*capacité*) – it interprets social domination as a system 'open' on all sides, into which elements from all social groups can be admitted. This interpretation, obscuring the real state of affairs, has become a central part of the authoritarian theory.

Even in the year in which Pareto's sociology appeared, the concept of the open system of domination only applied to a thin upper layer of social reality. From the point of view of the economic base the system of domination had long ago become closed along class lines, and the circulation of élites as he described it was only a peripheral feature of the social mechanism. But this made it all the easier for the ruling groups to adopt the theory of élites: against the firm background of the class-hierarchy a gentle circulation of élites was quite permissible; the economic and political apparatus was strong enough to regulate it within certain limits. What Pareto gave to the political disciples of his theory was above all the ability to grasp the central importance of certain psychological constants and mechanisms and to see the value of irrational, 'non-logical' actions for the stabilization of social domination. 'Ruling classes, like other social groups, perform both logical and non-logical actions, and the chief element in what happens is in fact the order, or system, not the conscious

3. Op. cit., para. 2054.   4. Op. cit., para. 2057.

will of individuals, who indeed may in certain cases be carried by the system to points where they would never have gone of deliberate choice.'[5] Pareto is the first to grasp and deal with the *psychological* problem of class domination in the monopolistic phase of capitalism; he is also the first to introduce authority into this social context.

It is the 'residues' which determine the organization of society; but the rationalized form of that organization is determined by the 'derivations' plus the 'appetites and interests' of which the 'residues' are the expression.[6] 'Residues' are certain socially effective psychological constants, which 'correspond' to certain simple instincts (appetites, tastes, inclinations) and interests possessed by men[7] and which constitute the real core of the 'non-logical actions' which are socially so relevant. The derivations can be described more or less as the rationalizations of the residues; they draw all their social strength from the residues which they transform into firm complexes of ideas.[8] If the residues are a 'manifestation of the emotions', the derivations are a 'manifestation of the need to reason'.[9] They function primarily for the maintenance of the 'social balance', or more concretely (as Pareto once says with regard to the social sciences): 'to persuade men to act in a certain manner considered useful to society.'[10]

The decisive feature is that these psychological constants and their rationalizations are now built into a theory of social domination. The stability and continuity of domination depend on the existence and effect of the 'residues' and 'derivations', and the particular proportion existing between the two elements. It is true that all domination rests on force and on the rationalization

---

5. Op. cit., para. 2254. 6. Op. cit., paras 861, 2205.
7. Op. cit., paras 850, 851. 8. Op. cit., para. 1397. 9. Op. cit., para. 1401.
10. Op. cit., para. 1403. For the sake of clarification, we shall quote the general division into *residues* and *derivations* in Pareto (paras 888, 1419). RESIDUES: (1) Instinct for combinations; (2) Persistence of aggregates (and particularly religious and family feelings); (3) Need to express sentiments by external acts; (4) Residues connected with sociality (particularly the need for uniformity; pity and cruelty and the sentiments of social rank); (5) Integrity of the individual and his appurtenances; (6) The sex residue. DERIVATIONS: (1) Assertion; (2) Authority; (3) Accords with sentiments or principles; (4) Verbal proofs.

of force, but these can never on their own guarantee the stability and continuity of domination: the more or less voluntary consent (*consentement*) of the dominated is required: 'everywhere there is a governing class which is small in numbers and which maintains itself in power partly by force, and partly with the consent of the governed class, which is much more numerous.'[11] And this consent rests basically on the presence of the residues and the derivations in the right proportions and on the ability of the governing class to employ them as a 'means of government'. Pareto elaborated the ideological character of these means of domination, pointing out that their social value derives not from their truth content but from their 'social usefulness' in obscuring the real background to social organizations and evoking 'sentiments' which provide a psychological anchorage for and perpetually reproduce the existing structure of domination. 'To sum up, these derivatives express above all the feeling of those who are firmly in possession of power and wish to retain it, and also the much more general feeling of the usefulness of social stability.'[12] They serve 'to calm' the governed: it is impressed upon them that all power comes from God, that any rebellion is a crime and that to achieve what is just only 'reason' and never 'force' may be used. 'This derivative has the main aim of preventing the governed from giving battle on a terrain which is favourable to them.'[13] But all derivations are in turn dependent on the psychological constants which lie deeper down in the layer of the subconscious and the irrational: '. . . the policies of governments are the more effective, the more adept they are at utilizing existing residues.'[14] Pareto recognizes that the relatively slow change in these psychological constants, and their resistance to the more rapid upheavals of social phenomena, are of decisive importance for the continuity of the social life-process: 'it is that also which assures continuity in the history of human societies, since the category (a) [the residues] varies slightly or slowly.'[15]

This also gives us our definition of the authority problem. It

---

11. Op. cit., paras 2244, 2251.   12. Op. cit., para. 2184.
13. Op. cit., para. 2192.   14. Op. cit., para. 2247.   15. Op. cit., para. 2206.

appears firstly as derivation, in its rationalized, manifest shape, and secondly as residue: as the feeling which underlies this manifestation. Under the heading of derivation Pareto is really only describing various relationships of authority;[16] he points to the particular 'pertinacity' of the phenomenon of authority: 'the residue of authority comes down across the centuries without losing any of its vigour.'[17] More important are the residues of which the authority relationship is the derivation: as its psychological basis we must consider above all the class of sentiments grouped under the heading 'persistence of aggregates'.[18] Once again those sentiments among them which have their roots in the family are in the foreground: relationships of family and kindred groups, relations between the living and the dead, relations between a dead person and the things that belonged to him in life, etc. Pareto saw the importance of the family in the preparation, maintenance and transmission of authority; on several occasions he emphasized that any weakening of this persistence of aggregates would directly threaten the stability of social domination. The second psychological anchorage of authority he sees in the sentiments of inferiors: subordination, affection, reverence, fear. 'The existence of these sentiments is an indispensable condition for the constitution of animal societies, for the domestication of animals, for the ordering of human societies.'[19] Here too Pareto gives a 'value-free' description of the phenomena, but the social function of the phenomena described becomes clearly evident precisely though this open description, which foregoes any moral or intuitive concepts and focuses completely on the usefulness of the psychological constants and mechanisms as a means of government. Much more clearly, indeed, than in Sorel, who at some points preceded Pareto in the discovery of unconscious psychological realms as the ground for social stabilization.

Above all, Sorel drew attention to the role of the family in the realization of social '*valeurs de vertu*'. The family is the 'mysterious

16. Op. cit., paras 1434–63.
17. Op. cit., para. 1439.
18. Op. cit., para. 1434.   19. Op. cit., para. 1150.

region . . . whose organization influences all social relations';[20] in it the values most prized by current society are realized, as for example, 'respect for the human person, sexual fidelity, and devotion to the weak'.[21] But, in contrast to Pareto, Sorel gives the family a moral and sentimental consecration: he praises the monogamous family as the 'administrator of the morality of mankind' without recognizing its connection with bourgeois society. Owing to his use of an intuitionist method, with its tendency towards making a general survey of the whole rather than dissecting it analytically, Sorel here completely misses the dialectical character of social objects. He sees the family statically, in the manner of either-or, and he has the same manner of viewing authority. His only way, beyond the alternatives of authority in the class state and lack of authority in anarchy, is to escape into metaphysical-moral dimensions.

Pareto's positivist analysis has a much greater affinity to the dialectics of social reality. It also allows him to reveal the double-edged character of the authority relationship which behind the backs of the bearers of authority, as it were, works also in the interests of those subject to authority. 'Nor can it be said that the subject class is necessarily harmed when a governing class works for a result that will be advantageous to itself regardless of whether it will be beneficial, or the reverse, to the former. In fact there are very numerous cases where a governing class working for its own exclusive advantage has further promoted the welfare of a subject class.'[22]

Pareto did not investigate the dynamic of the double-edged character of this relationship any further; he mechanically placed the positive and the negative element side by side. However, it is this dynamic which characterizes history.

20. *Reflections on Violence*, p. 180.   21. Op. cit., p. 261.
22. Pareto, op. cit., para. 2249.